Practical Time Series

Gareth Janacek

University of East Anglia
UK

ARNOLD

A member of the Hodder Headline Group
LONDON
Co-published in the United States of America
by Oxford University Press Inc., New York

First published in Great Britain in 2001 by
Arnold, a member of the Hodder Headline Group,
338 Euston Road, London NW1 3BH

http://www.arnoldpublishers.com

Co-published in the United States of America by
Oxford University Press Inc.,
198 Madison Avenue,
New York, NY 10016

Whilst the advice and information in this book are believed to be true and accurate at the date of going
to press, neither the author nor the publisher can accept any legal responsibility or liability for any
errors or omissions that may be made.

British Library Cataloguing in Publication Data
A catalogue record for this book is available from the British Library

Library of Congress Cataloging-in-Publication Data
A catalog record for this book is available from the Library of Congress

ISBN 0 340 71999 0

1 2 3 4 5 6 7 8 9 10

Commissioning Editor: Liz Gooster
Production Editor: Rada Radojicic
Production Controller: Martin Kerans
Cover Design: Terry Griffiths

Typeset in 10/12 pt Times by HK Typesetting Ltd, High Wycombe
Printed and bound in Great Britain by Redwood Books Ltd, Trowbridge

What do you think about this book? Or any other Arnold title?
Please send your comments to feedback.arnold@hodder.co.uk

Practical Time Series

Contents

Series preface

Arnold Texts in Statistics is a series that is designed to provide an introductory account of key subject areas in statistics. Each book will focus on a particular area, and subjects to be covered include regression analysis, time series, statistical inference and multivariate analysis. Texts in this series will combine theoretical development with practical examples. Indeed, a distinguishing feature of the texts is that they will be copiously illustrated with examples drawn from a range of applications. These illustrations will take full account of the widespread availability of statistical packages and other software for data analysis. The theoretical content of the texts will be sufficient for an appreciation of the techniques being presented, but mathematical detail will only be included when necessary. The texts are designed to be accessible to undergraduate and postgraduate students of statistics. In addition, they will enable statisticians and quantitative scientists working in research institutes, industry, government organizations, market research agencies, financial institutions, and so on, to update their knowledge in those areas of statistics that are of direct relevance to them.

David Collett
Series Editor

Preface

Yet another time series book?

There are many books on time series, some of which are excellent, so why another? This text is, in common with some others, aimed at a non-specialist audience. We expect both some statistical background and some tolerance of mathematics and our aim is to give a fairly broad view of time series. Our hope and expectation is that the reader will be able to *use* the methods described on their own data. As a consequence we take a pretty applied approach to the subject. It is important to have enough detail to understand what you are doing but it can be taken too far. To this end we have avoided much of the more complex and intricate mathematics and our position is helped by the fact that we supply software (see below). This means that the reader can, and we expect that they will, try the techniques we suggest on both the data we provide and on their own data.

The content of the book is pretty conventional. There is a fairly extensive coverage of ARIMA models because they are widely used. We also discuss state space models and the Kalman filter; indeed our likelihood estimation routine is based on this approach. As an example of our intent we provide a basic outline of the estimation problem and a means to do it – we do not give chapter and verse. It is our view that this is too specialist and that readers will be more interested in actually estimating parameters rather than reading about the details.

We also cover the frequency domain including concepts such as the spectrum and filtering. While these are quite complex mathematically the ideas are quite simple and they do give real insight into time series. Again the aim is basic concepts and their application. We feel strongly that if one can actually try the methods the mathematics is less intimidating and the ideas are more obvious.

Computation

Many time series methods have only become feasible with the advent of high speed computation. Fortunately, computing power is now cheap and our remaining problem is software. We have attempted a hands-on approach where the reader is encouraged to try things out and it follows that software is required.

One problem is that the general purpose packages tend to have rather limited focus, certainly in relation to our aims, while the more specialist programs, while admirable, are rather expensive.

It became inevitable that we would have to supply the reader with sufficient code to run the examples in the text. There were several possible computing environments and we ended up having to choose between Xlisp-Stat and R. Both are available across platforms and gave the feature-rich environment we require. After some agonizing we have gone for R. This is a programming environment for data analysis and graphics which is not unlike Splus. It was developed by Robert Gentleman and Ross Ihaka and has the great virtue of being free in the GNU sense.

The appropriate flavour of R for your system is available from the master CRAN (The Comprehensive R Archive Network) website at

```
http://www.ci.tuwien.ac.at/R/contents.html
```

or one of its mirrors. In its state R did not provide time series routines and it is these that we supply. They are available together with documentation, pointers to CRAN, and so on at

```
http://www.mth.uea.ac.uk/h200/tsbook
```

The reader who points their browser at our web site will also find data sets and other information. Of course, if a reader has problems they are welcome to write to the author.

We make no claims that our routines are at the leading edge of computation – they are written to be simple and to do useful analysis on our data sets and those of the reader. They are not designed for large data analysis projects although they may work on them. If any reader has alternatives or additions and would care to contact the author we will gladly mount the new material.

While we have pushed the computation into the background in the text we do urge the reader to try the methods described. The data are available at

```
http://www.mth.uea.ac.uk/h200/tsbook/data
```

as is the code at

```
http://www.mth.uea.ac.uk/h200/tsbook/code
```

Running the examples in the book before migrating to your own data is a sensible idea and we urge this course of action on the reader.

For example, if we take Ozone concentration, AZUSA (Arizona USA), from 1956 to 1970 by months (one of the data sets provided) then typing

```
tsplot(x)
```

in R gives a plot as shown in the following figure.

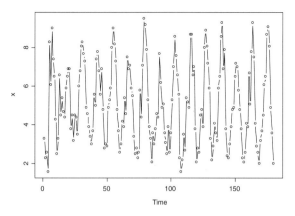

Fig. 1 *A time series*

Other commands give other options, so a running median of length 9 is computed by

msmooth(x,9)

is plotted against the original data in the following figure.

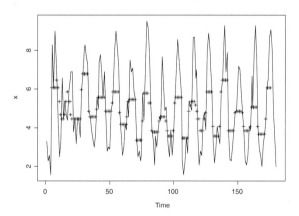

Fig. 2 *A time series with running median length 9*

Of course we would be glad to hear of any errors or omissions.

Anyone who writes a book must be aware of the debt that they owe to others. I owe a great deal to my teachers, colleagues and students. The authors of R and the development deserve great praise in bringing their wonderful statistical computing to the many.

Of course none of this would be possible without the support of my long suffering wife and family.

Glossary of symbols

$\hat{}$ Denotes an estimate, so $\hat{g}$ is an estimate of g.

α A smoothing parameter in exponential smoothing. Also the size of a statistical test.

α_t A vector of state space variables at time t.

β A smoothing parameter in exponential smoothing.

γ A smoothing parameter in exponential smoothing.

$\gamma(k)$ The autocovariance at lag k, that is $E[(X_t - \mu)(X_{t+k} - \mu)]$. Note this is sometimes written as $\gamma_{xx}(k)$ when dealing with more than one series to make it clear we are discussing the autocovariance for series X_t

$\gamma_{xy}(k)$ The crosscovariance at lag k between X_t and Y_t, that is $E[(X_t - \mu)(Y_{t+k} - \mu)]$.

$\Gamma()$ The transfer function of a filter.

$\delta, \delta()$ A small quantity.

δ_t A noise process. probably white.

$\boldsymbol{\delta}_t$ A vector of noise variables at time t.

Δ Sometimes used to denote $(1 - B)$. ∇ is more usual.

∇ $1 - B$

ϵ A small quantity.

ϵ_t A white noise process.

$\boldsymbol{\epsilon}_t$ A vector of noise variables at time t.

η_t A noise series

$\boldsymbol{\eta}_t$ A vector of noise variables at time t.

λ Usually a scale parameter in kernel smoothing.

μ The mean.

μ_t The mean at time t

ω Angular frequency (radians per unit time)

ϕ_k Coefficient in a polynomial, usually the AR polynomial in the backshift operator B.

$\phi(B)$ Usually the AR polynomial in the backshift operator B $\phi(B) = 1 - \sum_{k=1}^{p} \phi_k B^k$

$\boldsymbol{\phi}(B)$ Usually the AR polynomial in the backshift operator B for a vector model

ϕ, ϕ_t The transition matrix for updating state variables in a state space model.

$\psi(B)$ Usually a polynomial in the backshift operator B

$\psi(B)$ Usually the AR polynomial in the backshift operator B for a vector model
$$\psi(B) = 1 - \sum_{k=1}^{p} \psi_k B^k$$

π Well known mathematical constant $3.14159\ldots$

Ψ_k Vector of coefficients in the innovations for of a State space model

$\rho(k)$ The autocorrelation at lag k, that is $\gamma(k)/\gamma(0)$

$\rho_{xy}(k)$ The crosscorrelation at lag k between X_t and Y_t, that is $E[(X_t - \mu)(Y_{t+k} - \mu)]/(\sigma_x \sigma_y)$.

σ A standard deviation

τ Sometimes used as a symbol for a time lag

θ_k Coefficient of a polynomial in the backshift operator. Usually the one connected with the moving average part of a model.

$\theta(B)$ A polynomial in the backshift operator for the moving average component of an ARMA/ARIMA model. $\theta(B) = 1 + \sum_{k=1}^{q} \theta_k$

$\boldsymbol{\theta}(B)$ A polynomial in the backshift operator for the moving average component of vector ARMA/ARIMA model.

ω Angular frequency in radians per unit time. Note $\omega = 2\pi f$ where f is the frequency in cycle per unit time.

ξ_t A noise series

ζ_t A noise series

B The backshift operator $BX_t = X_{t-1}$

d The number of times to difference a series

D The number of times to seasonally difference a series

e_t Forecast error at time t

f Probability density function. Sometimes used to denote the power spectrum of the normalized power spectrum. The frequency in cycle per unit time. $f = \omega/2\pi$

$h(\omega)$ The power spectrum at frequency ω

g_u Coefficients in the model $X_t = \sum_{u=0}^{\infty} g_u \epsilon_{t-u}$. Note $g_0 = 1$

$G(x)$ $G(x) = \sum_{u=0}^{\infty} g_u x^{t-u}$. Note $g_0 = 1$

$I_N(\omega)$ The periodogram at frequency ω. $I_N(\omega) = \frac{1}{2\pi N} |\sum_{k=1}^{N} x_t e^{-i\omega t}|^2$

$J_N(\omega)$ The discrete fourier transform at frequency ω. $J_N(\omega) = \frac{1}{2\pi N} \sum_{k=1}^{N} x_t e^{-i\omega t}$

N The length of the observed series

p The order of the autoregressive part of a model

P The order of the seasonal autoregressive part of a model

q The order of the moving average part of a model

Q The order of the seasonal moving average part of a model

$\mathbf{R}$ The name of a computer language.

r_t Residual at time t

s A seasonal period

S_t A seasonal effect at time t.

t Time

T_t Trend at time t

x_t The observation at time t.

X_t A random variable indexed by time.

$\hat{X}_{t+k|t}$ The forecast of X_{t+k} made at time t.

$Z(\omega)$ A mathematical description of a series which is used in deriving properties of the power spectrum. Notably

$$E[dZ(\omega)\bar{d}Z(\theta)] = \begin{cases} f(\omega) & \text{if } \omega = \theta \\ 0 & \text{otherwise} \end{cases}$$

1
Introduction

A time series is a sequence of values $\{x_1, x_2, \ldots, x_t, \ldots\}$ observed through time. For most of this book we can happily assume that the observations are made at integer times $\{1, 2, \ldots, t, \ldots\}$ but in some cases we need other time points so you will often see references to the set of times called the index set T. For continuous time series, when t is continuous, there are complications that we will discuss later, see Chapter 6. Some examples of real time series are shown in Figure 1.1

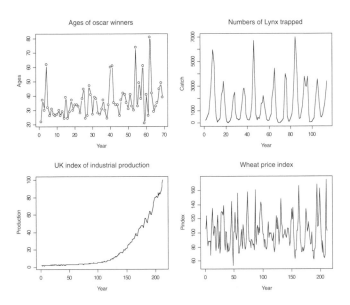

Fig. 1.1 *Some time series-ages of female oscar winners, -numbers of lynx trapped, -index of industrial production, -wheat price index*

Time series are ubiquitous; they arise in almost all situations where we keep records or make measurements. Our interest in them differs across applications, thus in some situations our main aim may be to forecast while in others the structure of the series or its interrelation with other series may be our main interest. To look at series in any useful way we need to think in terms of a model. This may be a model which reflects some knowledge of the generating mechanism, for example fish populations over time, or more usually a model which is capable of generating the behaviour observed.

As is usual in statistics we regard the observation at any time t, say x_t, as being the outcome of a random variable X_t. This means that we can think of a sequence of random variables $X_1, X_2, \ldots, X_t, \ldots$. The observed sequence is then just one outcome or realization of the random process $\{X_t\}$. This is the same process we usually follow in statistics – typically we have a set of random variables and the sample is a set of outcomes. The difference in this case is that for time series we usually do not have independence.

1.1 Smoothing

Before considering the structure of time series in any detail we look at a set of techniques for smoothing series. Some series may have an irregular appearance which we would like to smooth so as to get a clearer idea of the underlying structure. We may even go further and require the elimination of some characteristic which we regard as irrelevant. The traditional smoothing approach is via a simple model and this is where we begin.

1.1.1 Classical decomposition

Suppose we regard X_t as having the form

$$X_t = T(t) + C(t) + R(t) \tag{1.1}$$

where

- $T(t)$ is a trend component
- $C(t)$ is a cyclic or seasonal component
- $R(t)$ is the random effect

This mimics many of the observed properties of time series. Thus sales may increase over time, the increase being the trend $T(t)$ but in addition there may be a seasonal rise in the winter, indicated by $C(t)$. This cyclic term $C(t)$ repeats after a fixed length, s say, so

$$\cdots = C(t - s) = C(t) = C(t + s) = C(t + 2s) = \cdots \tag{1.2}$$

For example, a winter sales increase repeats every 12 months or every fourth quarter. The random effect is the unexplainable bit which we (hope) is purely random.

We begin with the aim of eliminating the seasonal effect. A simple way of doing this is to look at the seasonal averages. Thus if we have a cycle due to seasonal effects in monthly data it will be lost in the annual means. This is fine but does mean that we lose some data. A rather better idea is to take the average of the first s values, drop the first and take the next s and so on. This gives a sequence of new values, say x_t^*.

Thus

$$\frac{1}{s}\{x_1 + x_2 + \cdots + x_s\} = x_{\frac{s+1}{2}}^*$$

$$\frac{1}{s}\{x_2 + x_3 + \cdots + x_{s+1}\} = x_{\frac{s+1}{2}+1}^*$$

$$\vdots$$

$$\frac{1}{s}\{x_{N-s+1} + x_{N-s+2} + \cdots + x_N\} = x_{N-\frac{s+1}{2}+1}^*$$

You will notice that we have placed the new 'smoothed' values at the average time point. Surprisingly, if the original series has a cycle of period s the new series will have the cycle removed. In fact it is not difficult to show that if we assume that $C(t)$ is periodic with period s and that $\sum_{j=1}^{s} C(j) = 0$ then $\sum_{j=1}^{s} C(t+j) = 0$. If the trend $T(t)$ is approximately linear, say $T(t) = a + bt$ then $\frac{1}{s}\sum_{j=1}^{s} T(t+ j) = a + b(t + \frac{s+1}{2})$ so that the trend estimate is unbiased.

Once we have the trend we can estimate the seasonal effects by computing the difference between the original series and the smoothed one. If our moving average series is $\hat{X}_t$ then, provided the smoothing does not distort the trend,

$$\hat{X}_t = T(t) + C(t)$$

so the difference between the series and the moving average, $X_t - \hat{X}_t$ is the seasonal component. We can now estimate the effect of the first on the time point by averaging the appropriate differences. So we estimate the December effect by averaging the effects for all Decembers.

The reader may have noticed a problem here. If our period s is even then the smoothed series will be at time points corresponding to the midpoints of out measured time values. This is an irritation which can be compensated for in a rather simple way. When the smoothed series falls on these half time points we apply a new 2 points moving average to the smoothed series. This then brings the times back into line. The example below will bring out the computations involved.

Example 1.1. While in practice we would tend to use a suitable program it is helpful to see the calculations laid out for a simple case. In Table 1.1 we see a 4 point moving average being applied to a data series x_t (the second column). The resulting values are not at our chosen time points so to adjust a 2 point average is then applied to the 4 point one. This new series in column 4 gives us the smoothed value of the series we require. As we can see from Figure 1.2 we appear to have removed the cyclic effect.

Table 1.1 *An example using a 4 point and a 2 point MA*

t	x_t	4 point MA	2 point MA	Difference
1	3.3602			
2	−3.1769			
		2.0002		
3	0.3484		2.1422	−1.7938
		2.2842		
4	7.469		2.6236	4.8454
		2.9629		
5	4.4963		3.0096	1.4867
		3.0563		
6	−0.4621		2.9912	−3.4533
		2.9261		
7	0.7218		3.0187	−2.2969
		3.1114		
8	6.9484		3.5347	3.4137
		3.9579		
9	5.2374		4.4553	0.7821
		4.9527		
10	2.9242		5.494	−2.5698
		6.0354		
11	4.7006		6.0262	−1.3256
		6.0169		
12	11.2793		5.8444	5.4349
		5.6719		
13	5.1637		6.5995	−1.4358
		7.527		
14	1.5441		7.3245	−5.7804
		7.1219		
15	12.121		7.488	4.633
		7.854		
16	9.6588		8.1567	1.5021
		8.4593		
17	8.0922		8.3714	−0.2792
		8.2835		
18	3.9653		8.7272	−4.7619
		9.171		
19	11.4177			
20	13.2088			

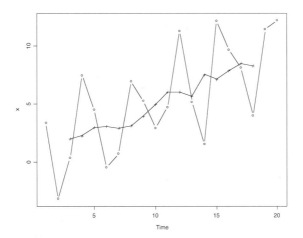

Fig. 1.2 *Smoothed series after 4 and 2 point MAs*

As we said, the residuals, or the difference between the original and smoothed values can be used for estimating the seasonals. If we take the average of all the first quarter residuals we have 0.138 45 which is our estimate of the first seasonal effect. In the same way we can calculate the remaining three seasonal effects. Conventionally the seasonal effects are assumed to sum to zero so we adjust them by subtracting a quarter of their sum from each value as shown in Table 1.2. Once we have a decomposition we can extrapolate the individual components and use these to make forecasts using our estimates in the model equation 1.1. We predict the trend using regression, this gives

$$T(t) = 0.0481 + 0.5146t$$

and hence we compute

$$T(21) = 10.8537, T(22) = 111.3683, T(23) = 11.8828, T(24) = 12.3974$$

We can now add seasonal terms to these trend predictions to give the forecasts.

This simple technique can work very well and has the advantage of being simple and robust. It is important that the trend effects are (locally) linear otherwise we may distort the trend estimates. There are other types of series where we need to be careful – in the calculations above we have made three assumptions

1. the model is of the form given by equation 1.1
2. we can assess, s, the period
3. the trend is locally linear

Table 1.2 *Seasonal estimates*

Seasonal	Raw value	Adjusted
c1	0.13845	0.238375
c2	−4.14135	−4.041425
c3	−0.195825	−0.0959
c4	3.799025	3.89895
sum	−0.3997	0

In many cases the model assumption may be unrealistic and it may be more plausible to assume that

$$X_t = T(t)C(t)R(t) \qquad (1.3)$$

This may be a more realistic model in cases where the amplitude of the seasonal cycles increases or decreases with the trend. One simple check is to overplot segments of the original series over the cycle.

Example 1.2. The series in Table 1.3 was analysed by Chatfield and Prothero (1973).

Table 1.3 *Monthly sales data from Chatfield and Prothero*

154	96	73	49	36	59	95	169	219	278	298	245
200	118	90	79	78	91	167	169	289	347	375	203
223	104	107	85	75	99	135	211	335	460	488	326
346	261	224	141	248	145	223	272	445	560	612	467
518	404	300	210	196	186	247	343	464	680	711	610
613	392	273	322	189	257	324	404	677	858	895	664
628	308	324	248	272							

There seems to be some increasing trend in the series and an increasing amplitude of variation. A useful check which should bring out any such effect is to 'overplot' months. This useful trick consists of plotting the series against the months $1, 2 \ldots 12$. A stable series will have a compact set of points for each month while variation or non stationarity will be highlighted. In Figure 1.3 we see that the amplitude of the cycles is increasing.

This is evident from the plot of the series but the overplot Figure 1.4 is rather more convincing. This overplot figure also gives us a chance to see the range of values for each month. If you have the software a series of boxplots is rather good as in Figure 1.5. From these figures it seems that we can either attempt a

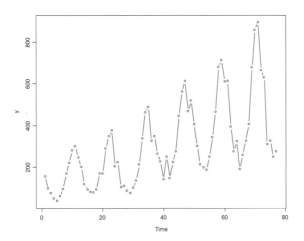

Fig. 1.3 *Chatfield Series*

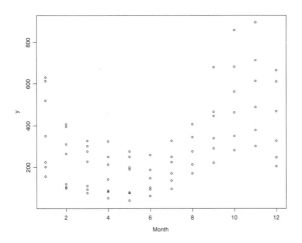

Fig. 1.4 *Overplot of Chatfield Series*

multiplicative model or take logs. If we take logs we get a more stable pattern (see Figure 1.6) and we can then smooth assuming that the logged series is additive.

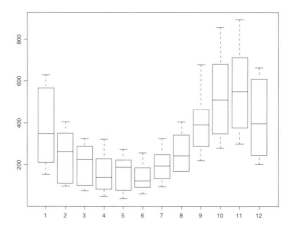

Fig. 1.5 *Overplot of Chatfield Series using boxplots*

Taking logs and then using a 12 point followed by a 2 point moving average seems adequate as can be seen in Figure 1.7.

1.1.2 Notation

We need some notation for moving averages. If we write the original 4 point moving average as the series of weights multiplying the terms in the original series we have $\frac{1}{4}[1, 1, 1, 1]$. This notation clearly extends to other moving averages of different lengths and with possibly unequal weights. In the example above we can write the operation of applying the moving averages as

$$\frac{1}{4}[1, 1, 1, 1]\frac{1}{2}[1, 1]x_t$$

using multiplication as a device for expressing a sequential application. It is not difficult to see that the two smoothing operations are equivalent to the five point MA $\frac{1}{8}[1, 2, 2, 2, 1]$.

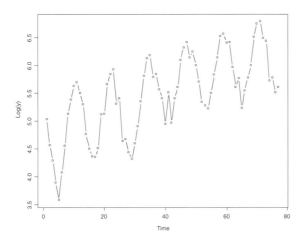

Fig. 1.6 *Log of Chatfield series*

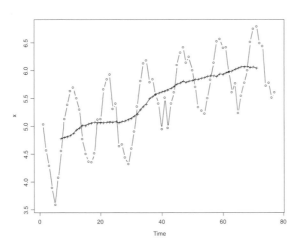

Fig. 1.7 *Smoothed log of Chatfield series*

1.2 Regression smoothers

The methods above can be recast in the form of a regression model. Rather than 1.1 we might consider

$$X_t = \beta_0 + \beta_1 t + \beta_2 q_1 + \beta_3 q_2 + \beta_4 q_3 + \beta_5 q_4 + \epsilon_t \qquad (1.4)$$

where q_j is zero when it is not quarter j and is one otherwise. As usual in regression models we assume that the ϵ_t are uncorrelated random variables with zero mean and common variance, say σ^2. This model is not quite right as it has too many parameters and if we try to fit a regression it will not work. We have to drop the constant term β_0 or one of the dummy variables $q_j, j = 1, \ldots 4$. Thus we could drop q_1 giving

$$X_t = \beta_0 + \beta_1 t + \beta_3 q_2 + \beta_4 q_3 + \beta_5 q_4 + \epsilon_t \qquad (1.5)$$

which is setting the first quarter as the benchmark for comparison. The reader with experience of linear models will recognize this as a standard procedure when one has a factor such as seasonality and a covariate like time. The results one obtains are much the same as for the widowed technique above. Regression (using R) gives for our quarterly example above

```
Coefficients:
              Estimate    Std. Error    t value    Pr(>|t|)
(Intercept)   1.09903     1.16837       0.941      0.36178
time          0.46344     0.08082       5.734      3.95e-05
q2           -4.77448     1.29560      -3.685      0.00221
q3           -0.33493     1.30314      -0.257      0.80066
q4            3.05259     1.31562       2.320      0.03483
```

```
Residual standard error: 2.045 on 15 degrees of freedom
Multiple R-Squared: 0.8404,   Adjusted R-squared: 0.7978
F-statistic: 19.75 on 4 and 15 degrees of freedom,
p-value: 7.695e-06
```

The experienced regression user will recognize the usual statistics such as R^2 but the whole point is that this technique gives us an explicit model and *all the advantages of regression*. The trend is immediate, it is just the time coefficient!

The drawback is that we assume a constant model over the whole time range. This may not be true and one of the strengths of the moving average approach is that it can cope with a smoothly varying trend. Of course with regression we have the option of using trigonometric terms as the explanatory variates. This gives a topic we will discuss further when we come to the periodogram. The log of the Chatfield data is shown in Example 1.3.

Example 1.3.

```
Coefficients:
              Estimate     Std. Error    t value    Pr(>|t|)
(Intercept)   5.0188536    0.0753052     66.647     < 2e-16
time          0.0214972    0.0009153     23.487     < 2e-16
   q2        -0.5147887    0.0951251     -5.412     9.97e-07
   q3        -0.7188543    0.0951383     -7.556     1.96e-10
   q4        -0.9804908    0.0951603    -10.304     3.11e-15
   q5        -1.1315845    0.0951911    -11.888     < 2e-16
   q6        -0.9702421    0.0990089     -9.800     2.33e-14
   q7        -0.6018434    0.0990047     -6.079     7.42e-08
   q8        -0.3263150    0.0990089     -3.296     0.001603
   q9         0.0733031    0.0990216      0.740     0.461841
   q10        0.3233911    0.0990428      3.265     0.001759
   q11        0.3655200    0.0990724      3.689     0.000466
   q12        0.0215009    0.0991104      0.217     0.828946
```

Residual standard error: 0.178 on 64 degrees of freedom
Multiple R-Squared: 0.9487,   Adjusted R-squared: 0.9391
F-statistic: 98.7 on 12 and 64 degrees of freedom,
    p-value: 0

Again we are providing the raw R output, hence the rather extravagant number of decimal places; three would surely be sufficient! Here qk is the effect of the month k. This is as we can see a pretty good fit and again we have separated out the trend. The fitted values are plotted for comparison in Figure 1.8

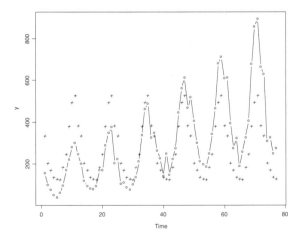

Fig. 1.8 *Smoothed log of Chatfield series*

1.3 Other moving averages

The moving averages that we developed above have equal, or mostly equal weights. However we can have other moving averages where the weights are quite radically different. Consider a regression problem where we decide to fit a cubic to 5 points. Perhaps on the

$$X_t = \beta_0 + \beta_1 t + \beta_2 t^2 + \beta_3 t^3 + \epsilon \tag{1.6}$$

If we use least squares as our estimation procedure we seek to minimize

$$Q = \sum_{t=1}^{5} \left(X_t - \beta_0 - \beta_1 t - \beta_2 t^2 - \beta_3 t^3 \right)^2 \tag{1.7}$$

We know that it is probably sensible to scale the time values of our variables so that they become $X_{-2}, X_{-1}, X_0, X_1, X_2$. In this case the least square criterion becomes

$$Q = \sum_{t=-2}^{2} \left(X_t - \beta_0 - \beta_1 t - \beta_2 t^2 - \beta_3 t^3 \right)^2 \tag{1.8}$$

If we differentiate with respect to the β coefficients to obtain the minimum we get the so called normal equations,

$$\sum_{t=-2}^{2} x_t = 5\beta_0 + \beta_1 \sum_{t=-2}^{2} t + \beta_2 \sum_{t=-2}^{2} t^2 + \beta_3 \sum_{t=-2}^{2} t^3$$

$$\sum_{t=-2}^{2} x_t t = \beta_0 \sum_{t=-2}^{2} t + \beta_1 \sum_{t=-2}^{2} t^2 + \beta_2 \sum_{t=-2}^{2} t^3 + \beta_3 \sum_{t=-2}^{2} t^4$$

$$\sum_{t=-2}^{2} x_t t^2 = \beta_0 \sum_{t=-2}^{2} t^2 + \beta_1 \sum_{t=-2}^{2} t^3 + \beta_2 \sum_{t=-2}^{2} t^4 + \beta_3 \sum_{t=-2}^{2} t^5$$

$$\sum_{t=-2}^{2} x_t t^3 = \beta_0 \sum_{t=-2}^{2} t^3 + \beta_1 \sum_{t=-2}^{2} t^4 + \beta_2 \sum_{t=-2}^{2} t^5 + \beta_3 \sum_{t=-2}^{2} t^6$$

Since

$$\sum_{t=-2}^{2} t = \sum_{t=-2}^{2} t^3 = \sum_{t=-2}^{2} t^5 = 0$$

While

$$\sum_{t=-2}^{2} t^2 = 10, \sum_{t=-2}^{2} t^4 = 34, \sum_{t=-2}^{2} t^6 = 130$$

and our equations become

$$\sum_{t=-2}^{2} x_t = 5\beta_0 + 10\beta_2$$

$$\sum_{t=-2}^{2} x_t t = 5\beta_1 + 34\beta_3$$

$$\sum_{t=-2}^{2} x_t t^2 = 10\beta_0 + 34\beta_2$$

$$\sum_{t=-2}^{2} x_t t^3 = 34\beta_1 + 130\beta_3$$

The midpoint of the polynomial is at β_0 which we can find by solving the equation. The solution is

$$\beta_0 = \frac{1}{35}[-3x_{-2} + 12x_{-1} + 17x_0 + 12x_1 - 3x_2]$$

which is just a moving average but with unequal weights! It follows that if we fit this polynomial in piecewise segments and use the constant term as the smoothed values this is equivalent to a moving average of length 5 with unequal weights. In fact, our moving average can be written in a fairly obvious way as a list of coefficients $\frac{1}{35}[-3, 12, 17, 12, 3]$. You will sometimes see this abbreviated to $\frac{1}{35}[-3, 12, \mathbf{17}]$. We can find moving averages for all such polynomials, thus for cubics

5 points $\frac{1}{35}[-3, 12, 17, 12, 3]$
7 points $\frac{1}{21}[-2, 3, 6, 7, 6, 3, -2]$
9 points $\frac{1}{231}[-21, 14, 39, 54, 59, 54, 39, 14, -21]$
11 points $\frac{1}{429}[-36, 9, 44, 69, 84, 89, 84, 69, 44, 9, -36]$
13 points $\frac{1}{143}[-11, 0, 9, 16, 21, 24, 25, 24, 21, 21, 16, 9, 0, -11]$
15 points $\frac{1}{1105}[-78, -13, 42, 87, 122, 147, 162, 167, 162, 147, 122, 87, 42, -13, -78]$

while similar expressions are available for quintic curves. The effect of the seven point is shown in Figure 1.9.

Such averages are useful in giving a way of smoothing series which requires very few assumptions and which is reasonably simple to use, especially with a spreadsheet. The drawbacks are precisely the same – the informal procedure does not lend itself to testing or model building. It is difficult to make a rational choice of polynomial and moving average length. Some theory is possible, see Anderson (1971), thus for a fixed length moving average the variance increases with polynomial order while for fixed order the bias decreases with length but this

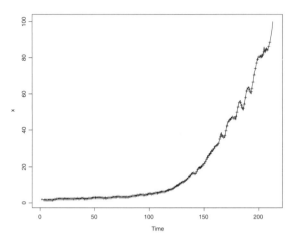

Fig. 1.9 $\frac{1}{21}[-2, 3, 6, 7, 6, 3, -2]$ *applied to UK index of industrial production*

result is of limited use. Also one needs to take care since cavalier use of successive moving averages may well lead to induced cyclic effects in the smoothed series. Moving averages have a long history linked to interpolation from tables. Interpolation formulae involve differences and smoothing formulae are used to give a graduation which has smooth successive differences. A moving average is said to be correct to order q if differences of this order polynomial remain unaffected. One popular example is Spencer's 15 point formulae which consist of applying $\frac{1}{4}[-3, 3, 4, 3, -3]$ then $\frac{1}{5}[1, 1, 1, 1, 1]$ then $\frac{1}{4}[-3, 3, 4, 3, -3]$ followed by $\frac{1}{4}[-3, 3, 4, 3, -3]$.

1.4 Operators and notation

All the moving averages we have seen can be thought of as weighted sums of terms like x_t and x_{t-1} or x_{t+1}. It is quite common to think of these values as x_t and time shifted values. We will make use of the shift operator B in later chapters. This is defined as

$$Bx_t = x_{t-1}$$

so that $B^2 x_t = x_{t-2}$ and $B^{-1} x_t = x_{t+1}$. This means we can write moving averages in terms of shifts so the 5 point average $\frac{1}{5}[1, 1, 1, 1, 1]$ is just $\frac{1}{5}(B^2 + B^1 + 1 + B + B^2)x_t$. It is quite common to see averages written in terms of these backshift operator B and the related difference operator $\nabla = (1 - B)$. This is not something we shall do but it is fairly straightforward if you have an algebra package such as Maple or Mathematica.

1.5   Other smoothers

There are many other smoothers available to time series analysts and we will take the time to look at one in this section. Smoothing is an area of considerable development and we will restrict ourselves to a brief overview.

We begin with a simple variant of the moving average process where we use running medians in place of means. We take a sequence of P observations and use the median as the middle value. So

$$\text{median}\left\{x_1, x_2, \ldots, x_p\right\} = x^*_{\frac{p+1}{2}}$$

$$\text{median}\left\{x_2, x_3 + \ldots, x_p\right\} = x^*_{\frac{p+1}{2}+1}$$

$$\vdots$$

$$\text{median}\left\{x_{N-p+1}, x_{N-p+2}, \ldots, x_N\right\} = x^*_{N-\frac{p+1}{2}+1}$$

Running medians of this form tend to give less smooth traces than their running mean counterparts. They are, as one might expect, very much less likely to be distorted by outliers in the series. This robust behaviour can be very useful since in many cases a simple smoother will reveal a few unusual observations. The motorcycle data from Hardle (1990) are smoothed by a 7 point median smoother in Figure 1.10. We have a rather rough smoothed series but the difference between the smoothed series and the original can be used to pick up outliers. Notice we have not mentioned a model and this is usual with medians. They are primarily used just as smoothers.

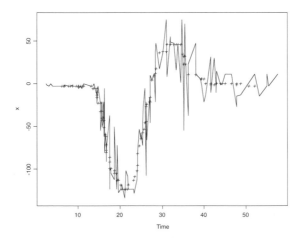

Fig. 1.10 *Median smoother applied to motor cycle data*

The interested reader may be interested in refinements on median smoothers proposed by Velleman (1980) and Mallows (1976). Another popular robust smoother is LOWESS (Locally weighted Scatter plot Smoothing) proposed by Cleveland. This begins with a local polynomial fit which is then modified by weighted residuals. Essentially one fits

$$\frac{1}{n} \sum_{i=1}^{n} W_{k,i}(t) \left[x_t - \sum_{j=0}^{p} \beta_j t^j \right]$$

where the weights $W_{k,i}$ depend on neighbouring points. The regression is then refitted but with weights modified by the residuals. The choice of p=1 is common giving the so called super smoother.

The details can be found in, for example, Hardle (1990). If you use R you will find that lowess is provided as a function in the form

```
lowess(x, y, f=2/3, iter=3, delta=.01*diff(range(x)))
```

As an example we tried this lowess smoother on the motor cycle data. From Figure 1.11 we see that the smoothing looks very reasonable and is certainly smoother than the median, as one might expect.

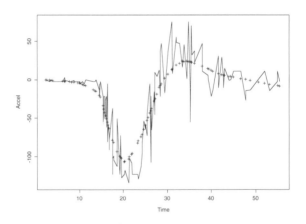

Fig. 1.11 *Motor bike series smoothed using the 'super smoother'*

1.6 Nearest neighbour and kernel smoothers

There has been a good deal of interest in smoothing in statistical circles, and there are some general smoothers which can be used in time series. In fact we can generalize very simply to bivariate data (x_i, y_i) $\quad i = 1, 2, \ldots, n$ where the y_i are the responses and we have some model of the form

$$y_i = f(x_i) + \epsilon_i \tag{1.9}$$

A simple approach is to average the responses in a local neighbourhood so our estimate of $f(x_i)$ is a weighted average of the y_i values near x_i. The size of the neighbourhood (the number of points included is known as the span or window size). A symmetric window for example could be

$$\max\left(1, i - \frac{m[wn] - 1}{2}\right), \ldots, \min\left(n, i + \frac{m[wn] - 1}{2}\right)$$

where w is some proportion lying in $[0, 1]$. It is clear that for narrow neighbourhoods we have rough estimates while for wide windows the resulting estimates are smoother. Near the endpoints asymmetry leads to bias.

The average can be an arithmetic mean, giving the running mean smoother or the more resistant running median we discussed above. Another simple form is the running line smoother where a linear form

$$\hat{f}(x_i) = \hat{\alpha}_i + \hat{\beta}_i x_i$$

is used where the constants are the usual least squares estimates. We have tacitly assumed that the smoothed response is needed at a point in the neighbourhood. If this is not true then we interpolate adjacent fits.

An alternative is to use varying asymmetric neighbourhoods. One common choice is the nearest neighbourhood. For k in the integers the k-nearest neighbourhood (k-NN) is a weighted average based on the neighbourhood of x defined by

$$\{i : x_i \text{ is one of the k-nearest observations to x}\}$$

Of course near has to be defined in terms of some distance measure $d(x, x_i)$ and the degree of smoothing is fixed by the value of k, the proportion of point in the neighbourhood being $w = k/n$. A linear k-NN estimate could be

$$\sum_{i=1}^{n} a(x, x_i) y_i$$

where the $a(x, x_i)$ are zero outside the neighbourhood and $1/k$ within it. A nice feature of these k-NN schemes is that they are easily modified for multivariate cases.

Kernel smoothing is based on the form

$$f_K(x) = \sum_{i=1}^{n} a(x, x_i) y_i$$

where the weights $a(x, x_i) i = 1, 2, \ldots, n$ at x are determined by a density function adjusted by a scale parameter, say λ. There are several widely used kernel functions including the standard Gaussian, the Epanechnikov

$$K(u) = \frac{3}{4}(1 - u^2) \text{ for } |u| \leq 1 \text{ and } 0 \text{ otherwise}$$

or the minimum variance kernel

$$K(u) = \frac{8}{4}(3 - 5u^2) \text{ for } |u| \leq 1 \text{ and } 0 \text{ otherwise}$$

The ideas of kernel smoothing follow from the application of kernel smoothing techniques in spectral analysis and density estimation. The idea is that to smooth a wiggly function we multiply by a weight function and integrate

$$f_{\text{smooth}}(x) = \int_{-b}^{b} f(y) K(x - y) \, dy$$

If the range of the integral is small then the smoothed value is close to the actual value as is the case if the kernel function is very peaked. The kernel smoother for time series, or in general for non-parametric regression is $\sum_{i=1}^{n} a(x, x_i) y_i$ where the weights $a(x, x_i)$ are chosen proportional to the kernel function $K(\frac{x - x_i}{\lambda})$. This implies that the weight will decrease at the rate of decay of the kernel, the scale parameter λ is however the dominant factor. In practice the commonly used form of the weight is the Nadaraya-Watson weight

$$a(x, x_i) = \frac{\frac{1}{n\lambda} K(\frac{x - x_i}{\lambda})}{\sum_{j=1}^{n} \frac{1}{n\lambda} K(\frac{x - x_j}{\lambda})}$$

There are other weight schemes and we refer the reader to Hardle (1990) for more details. While R has no built-in Kernel smoothers we provide a simple minded routine which is used to provide the plots. There are several much more sophisticated public available sets of functions and these can be found at the R web site.

As you will notice the kernel and k-NN smoothing procedure are really very closely related, especially with equally spaced x values. If we are merely using these schemes to smooth time series it is hard to decide which is the more appropriate. The nearest neighbour requires an averaging function and the choice of the k, while for kernel methods one must choose a window and a bandwidth. There is more literature on the kernel methods and we have rather more theory but we refer the interested reader to Hardle (1990). If we use a Gaussian kernel of width 7 on the motorcycle data we have the result in Figure 1.12

If we take the kernel width very much wider, say 50, we have an almost flat trace while at 25 we have some idea of the shape of the data. As with all smoothing the problem is to specify the minimum bandwidth or the width of the narrowest peak of interest in the data. This is a technical problem which is not really solved.

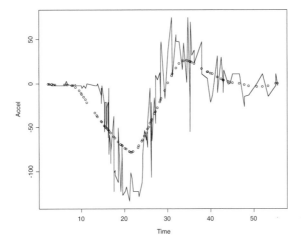

Fig. 1.12 *Motor bike series smoothed using a kernel smoother*

The best advice we can offer is to try several widths until one gives a reasonable smoothed finish.

The reader will appreciate that much of our discussion has been rather ad hoc. To some extent this is inevitable with smoothing. Very often we smooth to explore the data and some informality is inevitable. There is in the end a considerable subjective input.

1.7 Computation

The classical moving average smoother is easily computed by hand, or more easily using a spreadsheet programme. Most nearest neighbour smoothers need a spreadsheed programme as a minimum. We provide R functions for all the cases we describe (Lowess is a built in function) and we urge the reader to try them and to examine the effects of changing the lengths of the smoothers. It is much easier to do with direct graphical output!

In general do bear in mind that the degree of smoothness required depends on the application!

1.8 Exercises

1. The following data give the sales of a company for each quarter from 1987. Use a suitable moving average to decompose the series into trend and seasonal components. Hence forecast the series for 1991 and 1992.

Year	Quarter			
	1	2	3	4
1985	322	144	472	821
1986	408	247	626	925
1987	434	259	681	1277
1988	829	435	940	1639
1989	1222	592	1055	2000
1990	1278	768	1415	2417
1991	1260			

2. Smooth the following series. Is there a trend? Are there any seasonal effects?

Co2 (ppm) at Mauna Loa. Jan 1965-Jan 1980 monthly											
319.32	320.36	320.82	322.06	322.17	321.95	321.20	318.81	317.82	317.37	318.93	319.09
319.94	320.98	321.81	323.03	323.36	323.11	321.65	319.64	317.86	317.25	319.06	320.26
321.65	321.81	322.36	323.67	324.17	323.39	321.93	320.29	318.58	318.60	319.98	321.25
321.88	322.47	323.17	324.23	324.88	324.75	323.47	321.34	319.56	319.45	320.45	321.92
323.40	324.21	325.33	326.31	327.01	326.24	325.37	323.12	321.85	321.31	322.31	323.72
324.60	325.57	326.55	327.80	327.80	327.54	326.28	324.63	323.12	323.11	323.99	325.09
326.12	326.61	327.16	327.92	329.14	328.80	327.52	325.62	323.61	323.80	325.10	326.25
326.93	327.83	327.95	329.91	330.22	329.25	328.11	326.39	324.97	325.32	326.54	ʼ327.71
328.73	329.69	330.47	331.69	332.65	332.24	331.03	329.36	327.60	327.29	328.28	328.79
329.45	330.89	331.63	332.85	333.28	332.47	331.34	329.53	327.57	328.53	329.69	
330.45	330.97	331.64	332.87	333.61	333.55	331.90	330.05	328.58	328.31	329.41	330.63
331.63	332.46	333.36	334.45	334.82	334.32	333.05	330.87	329.24	328.87	330.18	331.50
332.81	333.23	334.55	335.82	336.44	335.99	334.65	332.41	331.32	330.73	332.05	333.53
334.66	335.07	336.33	337.39	337.65	337.57	336.25	334.39	332.44	332.25	333.59	334.76
335.89	336.44	337.63	338.54	339.06	338.95	337.41	335.71	333.68	333.69	335.05	336.53
337.81	338.16	339.88	340.57	341.19	340.87	339.25	337.19	335.49	336.63	337.74	338.36

3. Suppose we apply an m point equally weighted moving average to a series with a locally linear trend. Show that the smoothed series is a good approximation of the linear trend.

4. Write the moving average $\frac{1}{5}[1, 1, 1, 1, 1]$ in terms of the shift operator B. Express the moving average in terms of the difference operator $\nabla = 1 - B$.

5. Use a regression method to forecast the series in question 1.

6. Smooth the ecg data given in the databank

7. In January 1983 the UK government introduced legislation which made the wearing of seat belts compulsory for those in the front seats of cars. The aim of the law was to reduce the numbers killed and seriously injured in accidents. As there are now data available for before and after the event we may reasonably attempt to see what, if any, effect the law had on injuries. The obvious and simplest question is 'did the law have any effect?' The tables below show the numbers of accidents by months for periods 2 years before and 2 years after the introduction of the law. These data were analysed using rather sophisticated methods, however there is a lot we can do using basic statistical tools.

For each table compare the trends in accidents after removing the seasonal effects.

Number of car drivers killed or seriously injured			
Year	1981	1982	1983	1984
Month			
Jan	1474	1456	1494	1357
Feb	1458	1445	1057	1165
Mar	1542	1456	1218	1282
Apr	1404	1365	1168	1110
May	1522	1487	1236	1297
Jun	1385	1558	1076	1185
Jul	1641	1488	1174	1222
Aug	1510	1684	1139	1284
Sep	1681	1594	1427	1444
Oct	1938	1850	1487	1575
Nov	1868	1998	1483	1737
Dec	1726	2079	1513	1763

Number of car drivers killed			
Year	1981	1982	1983	1984
Month			
Jan	111	115	120	92
Feb	106	104	95	86
Mar	98	131	100	81
Apr	84	108	89	84
May	94	103	82	87
Jun	105	115	89	90
Jul	123	122	60	79
Aug	109	122	84	96
Sep	130	125	113	122
Oct	153	137	126	120
Nov	134	138	122	137
Dec	99	152	118	154

2
Exponential smoothing

Often one encounters situations where our main interest is forecasting rather than explicit model building and in these circumstances exponential smoothing is a useful and very effective technique. We begin with the assumption that recent events are rather more important than those in the distant past, the idea used by Brown (1959, 1963) when he proposed exponential smoothing. The method is simple, given a series x_t we construct new series

$$M_{t+1} = \alpha x_t + (1 - \alpha)M_t \qquad 0 \le \alpha \le 1$$

This simple recurrence gives

$$M_{t+1} = \alpha x_t + \alpha(1 - \alpha)x_{t-1} + \alpha(1 - \alpha)^2 x_t + \cdots + \alpha(1 - \alpha)^t x_0$$

Since α lies between 0 and 1 the coefficients $\alpha(1 - \alpha)^j$ are decreasing as j increases. This means that while the new series M_t depends on the whole x_t series the influence of past terms is decreasing. Brown suggested that the derived series M_t be used to forecast. If we write the k step ahead forecast made at time t as $\hat{X}_{t+k|t}$ we have

$$\hat{X}_{t+1|t} = \alpha x_t + (1 - \alpha)\hat{X}_{t|t-1} \tag{2.1}$$

It is easy to express this in terms of forecast errors

$$\hat{X}_{t+1|t} = \alpha[x_t - \hat{X}_{t|t-1}] + \hat{X}_{t|t-1} = \alpha e_t + \hat{X}_{t|t-1} \tag{2.2}$$

where $e_t = x_t - \hat{X}_{t|t-1}$ is the error in the forecast at time t.

The simplicity of this scheme, where the current forecast is a modification of the last one, is very attractive when one wishes to make many forecasts. For example, with a mail order catalogue there may be several thousand lines to forecast and a simple and computationally inexpensive procedure has many attractions.

Assuming that this simple idea works the first problem is the choice of α. If we make some simplifying assumptions – that the series has an infinite past and that

$$\operatorname{corr}(X_t, X_{t+k}) = \rho^k$$

We can show that if we wish to minimize the sum of squares

$$Q = \sum_{j=2}^{m} e_j^2$$

then we can show that

$$\alpha = \begin{cases} 0 & \text{if } -1 \le \rho \le \frac{1}{3} \\ 1 - (1 - \rho)/(2\rho) & \text{if } \frac{1}{3} < \rho \le 1 \end{cases}$$

For details see Gourieroux and Montfort (1997).

A more realistic process is to use a numerical minimization process to find suitable α. This is a fairly simple process. Given $x_1, x_2, \ldots, x_N$ we minimize

$$Q = \sum_{t=1}^{m} [x_{t+1} - \hat{X}_{t+1|t}]^2 = \sum_{t=1}^{m} \left[x_{t+1} - \alpha \sum_{j=0}^{t} (1 - \alpha)^j x_{t-j} \right]^2$$

For a series with no trends we can have a quite effective forecast and for some series, as we shall see in later chapters, these forecasts are optimal. In the minimization process it is not uncommon for the minimum of the sum of squares Q to occur at or near the boundaries of the allowable range for α. While such a value may be optimal it is often better to choose a less extreme α especially if the sum of squares surface is fairly flat near the minimum.

Monthly sales of jeans

We take the sales of jeans in the UK by month in Table 2.1 – read rowwise.

Table 2.1 *Sales of jeans in the UK by month*

1998	1968	1937	1827	2027
2286	2484	2266	2107	1690
1808	1927	1924	1959	1889
1819	1824	1979	1919	1845
1801	1799	1952	1956	1969
2044	2100	2103	2110	2375
2030	1744	1699	1591	1770
1950	2149	2200	2294	2146
2241	2369	2251	2126	2000
1759	1947	2135	2319	2352
2476	2296	2400	3126	2304
2190	2121	2032	2161	2289
2137	2130	2154	1831	1899
2117	2266	2176	2089	1817
2162	2267			

As a start up value for the smoothed series we take the average of the first three values in the sales series. The successive terms generated by exponential smoothing are shown in Table 2.2. When we plot the squared error, using 12 values of the series, we get Figure 2.1.

Table 2.2 *Some smoothed series using the jeans sales data*

series			
alpha	0.1	0.5	0.9
1998	1983	1983	1983
1968	1981.5	1975.5	1969.5
1937	1977.05	1959.25	1940.25
1827	1962.045	1902.025	1838.325
2027	1968.5405	1994.5225	2008.1325
2286	2000.28645	2127.27025	2258.21325
2484	2048.65781	2242.14323	2461.42133
2266	2070.39202	2157.3289	2285.54213
2107	2074.05282	2088.69601	2124.85421
1690	2035.64754	1882.02641	1733.48542

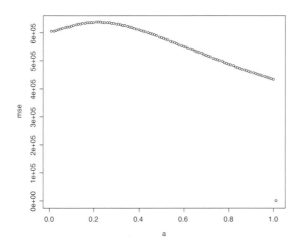

Fig. 2.1 *Mean square error of prediction for the jeans series using 12 values*

As is not uncommon there is no clear optimum value and an α close to 1 is indicated. The reader might like to plot the predicted against actual values for differing α values to see how the forecasts change with the mse.

As a contrast we also plot the mean square error of prediction (Figure 2.2) for the artificial series shown in Table 2.3. As we can see the minimum is around 0.18 which immediately gives us a value for α.

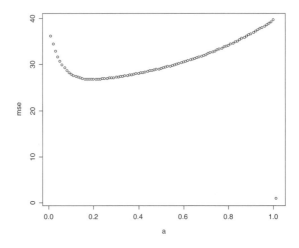

Fig. 2.2 *Mean square error of prediction for the artificial series using 18 values*

Table 2.3 *An artificial series*

0.100	0.100	0.036	1.472	0.077	0.871	2.811	2.544
0.438	2.064	−1.298	1.117	1.996	0.181	0.325	0.623
1.732	2.986	1.447	1.085	1.102	−0.644	2.247	0.029
0.396	0.022	0.956	1.889	0.121	−0.805	−0.13	0.841
−0.377	0.565	1.488	−0.681	0.819	−0.23	0.772	−1.389
1.09	1.225	1.196	0.762	0.215	1.71	−0.095	2.038
2.038	1.604	0.442	−0.374	1.887	−0.869	0.935	2.417
0.973	2.461	0.403	0.658	1.765	−0.131	−0.771	−0.911
0.883	2.155	−0.288	−0.312	1.038	0.433	−0.810	1.313
−0.219	1.532	1.136	−0.018	1.61	2.08	1.064	2.726
−0.649	1.621	1.396	0.459	1.405	−0.201	0.39	2.128
−1.576	0.93	1.501	1.282	3.101	0.726	0.046	1.192
0.283	0.42	1.047	1.299	2.083	−0.178		

2.1 Generalized exponential smoothing

As one might expect, simple exponential smoothing does not always work well and one common case which causes problems is when the series we wish to forecast has a trend. Applying the exponential forecasting procedure to a trending series it is not hard to see that there must be a *lag* in the forecast. If we consider the simple example $x_t = a + bt$ for constant coefficients

$$\hat{X}_{t+1|t} = \alpha \sum_{j=0}^{t} (1 - \alpha)^j [a + b(t - j)]$$

and assume that t is sufficiently large for

$$\sum_{j=0}^{t} (1 - \alpha)^j = \frac{1}{1 - (1 - \alpha)} = 1/\alpha$$

while

$$\sum_{j=0}^{t} j(1 - \alpha)^j = \frac{1}{[1 - (1 - \alpha)]^2} = \frac{1}{\alpha^2}$$

then

$$\hat{X}_{t+1|t} = a + bt - \frac{b}{\alpha}$$

and we can see that the forecasts lag the series.

Brown suggested double exponential smoothing as a way of getting around this problem but the more common approach is the one proposed by Holt (1957) and Winters (1960). Suppose we have a series but with a trend. We construct two series one for the level

$$M_{t+1} = \alpha x_t + (1 - \alpha)(M_t + T_t) \qquad 0 \le \alpha \le 1 \qquad (2.3)$$

and one for the trend

$$T_{t+1} = \beta(M_{t+1} - M_t) + (1 - \beta)T_t \qquad 0 \le \beta \le 1 \qquad (2.4)$$

the forecast at time t is then

$$\hat{X}_{t+k|t} = M_{t+k} + kT_t \qquad (2.5)$$

The explicit incorporation of the trend makes this a surprisingly effective technique when forecasts are required. For a good and comprehensive account see Chatfield (1978). As in the simple case we would use numerical techniques to find values of the parameters α and β which minimize

$$Q = \sum_{t=1}^{m} (x_{t+1} - \hat{X}_{t+1|t})^2$$

If we wish, we can go further and add a seasonal effect. Suppose we have a series with a seasonal effect C_t whose period is s, that is

$$\cdots = C_{t-s} = C_t = C_{t+s} = C_{t+2s} = \cdots$$

For an additive model we have three equations, one for the level

$$M_{t+1} = \alpha[x_t - C_{t-p}] + (1 - \alpha)(M_t + T_t) \qquad 0 \leq \alpha \leq 1 \qquad (2.6)$$

and one for the trend

$$T_{t+1} = \beta(M_{t+1} - M_t) + (1 - \beta)T_t \qquad 0 \leq \beta \leq 1 \qquad (2.7)$$

and one for the seasonal effect

$$C_{t+1} = \gamma(x_t - M_t) + (1 - \beta)C_{t-s} \qquad 0 \leq \gamma \leq 1 \qquad (2.8)$$

The forecast at time t for k steps ahead is just

$$\hat{X}_{t+k|t} = M_t + kT_t + C_{t+k-s} \qquad (2.9)$$

The reader will note that the form of the forecasting function assumes that the factors are additive. This may not be true – one often sees series where the seasonal cycles increase with the trend. A simple approach in this case is to take logs and to deal with the log series. Many authors do, however, explicitly modify their equations. For example, for the level

$$M_{t+1} = \alpha[x_t/C_{t-s}] + (1 - \alpha)(M_t + T_t) \qquad 0 \leq \alpha \leq 1 \qquad (2.10)$$

for the trend

$$T_{t+1} = \beta(M_{t+1} - M_t) + (1 - \beta)T_t \qquad 0 \leq \beta \leq 1 \qquad (2.11)$$

for the seasonal and one for the seasonal effect

$$C_{t+1} = \gamma(x_t/M_t) + (1 - \beta)C_{t-s} \qquad 0 \leq \gamma \leq 1 \qquad (2.12)$$

The forecast at time t is then

$$\hat{X}_{t+k|t} = [M_{t+1} + kT_{t+1}]C_{t+k-s} \qquad (2.13)$$

The choice of parameters is, once again, determined using a numerical minimization. We hope to have a long enough fitting section of our series to allow the estimates to be reasonably insensitive to the starting values.

We have not said very much about the choice of initial values. Of course our hope is that, whatever our choice after a few iterations, the start up effect will become negligible. This can of course be checked. We usually pick the mean of the first few observations to set the level, over a full cycle for seasonal data, and we average the first few values of the differences for the trend. Seasonals are more difficult, especially when we do not have many cycles to gauge the seasonal magnitudes. We can use zeros, but this may introduce a transient. It is well to remember that if we need a seasonal then we will need several cycles to gain information about the seasonal effects.

Table 2.4 *Shampoo sales from Hyndman*

266.0	145.9	183.1	119.3	180.3	168.5	231.8	224.5	192.8
122.9	336.5	185.9	194.3	149.5	210.1	273.3	191.4	287.0
303.6	289.9	421.6	264.5	342.3	226.0	339.7	440.4	315.9
439.3	401.3	437.4	575.5	407.6	682.0	475.3	581.3	646.9

2.1.1 Shampoo sales

As an example we consider the sales of shampoo from the Hyndman time series collection. The data are shown in Table 2.4 (read as rows)

Fitting a level and trend model, using 18 start up values, gives parameters $\alpha = 0.3$ and $\beta = 0.9$ using the R function provided to examine the surface, as seen in the plot in Figure 2.3. This is in fact log of the surface which gives a rather more useful contour plot.

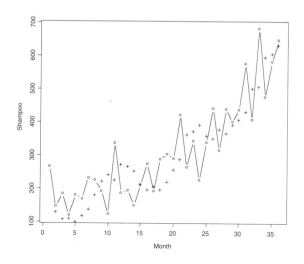

Fig. 2.3 *Prediction for the shampoo series using 18 values, + indicates prediction*

We did gloss over a problem in the discussion above. We do need to provide start up values for the mean level and trend terms. One possibility is to set these to zero with the expectation that these zero values will have a transient effect. In fact in the shampoo example we did not zero but some plausible values, $M_1 = (x_1 + x_2 + x_3 + x_4)/4$ to set the level and $T_1 = (x_4 - x_1)/3$ to start the trend may be preferable. We are fairly relaxed about these as estimates as we expect the effect of the start up values to be quickly lost. The reader might like to see if more refined estimates are any better – we provide the R function! Seasonals of course require care since long period effects persist by definition and may require quite good initial estimates.

It is often wise to look at the squared error of prediction over a range of values. Our minimization is a good one, but you only have our word. Using the supplied R function a contour plot of the error surface can be seen in Figure 2.4. The minimum is indicated and we can see that the function we are minimizing is quite flat around the minimum although to see this we need to remove the very high peaks at the edge of the plot.

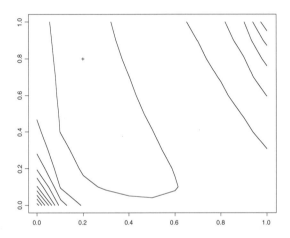

Fig. 2.4 *Sum of square surface*

These methods and their multiplicative variants are well used and by all reports have done well in various situations, in particular forecasting competitions. We will propose some reasons why this may be so when we consider ARIMA models. As with most forecasting methods one really has to try them out to see how they perform. Ideally one would estimate the optimum parameter values on a section of the series and then look at the performance of the potential forecasting function on the remainder. This is all rather subjective but there is no substitute to seeing how the fitted model works! We urge our reader to try some examples on several series with different training segments to estimate the discount parameters α, β, γ.

There are many variations on the theme of exponential smoothing. Many have been devised for specific circumstances but most are simple to compute and to understand. This is one of the real virtues of exponential smoothing. Our advice to the reader is to try it and see. We now turn our attention to rather more theoretical ideas.

2.1.2 Computation

There are several programmes which give the user the chance to optimize smoothing parameters. The smoothes series themselves are simple to generate using a spreadsheet and it is possible to do informal optimization this way.

We provide R functions using the built in minimizer

```
nlm
```

for two and three parameter smoothers. This appears to work quite well and it does give the reader a chance to try out their own ideas. You might like to see how many of your favourite series have smoothing parameters away from 0 and 1. For a single parameter it is simplest to look at the mean square error plot by eye.

2.2 Exercises

1. If we define the simple exponentially smoothed series M_t by $M_t = \alpha X_{t-1} + (1 - \alpha)M_{t-1}$ find the response of M_t to the following inputs

   (a) The step $X_t = 0$ for $t < 0$ $X_t = 0$ otherwise
   (b) The impulse $X_t = 1$ then $t = t_0$ and zero otherwise
   (c) The parabola $X_t = t^2$

2. The table below gives the annual prices of Minis from 1959 at constant (1959) prices. Construct an exponential forecasting scheme for this data set.

496.95	487.21	468.82	482.35	401.48	405	386.35	378.76	395.3	412.88
415.85	415.15	381.1	383.21	370.71	429.85	442.14	475.08	492.68	503.65
494.6	477.16	464.24	435.53	442.25	453.9	464	457.1	446.3	485.6

3. The table below gives the actual annual prices of Minis from 1959. Construct an exponential forecasting scheme for this data set.

496.95	496.95	496.95	526.25	447.65	469.8	469.8	478	508.75	561.1
595.5	638.5	640.63	695.15	738.84	1003.26	1299	1587	1893	2091
2404.5	2683.6	2899	2899	3098	3338	3598	3650	3725	4290

3
Stationary series

We have spent the last two chapters looking at informal and rather ad hoc methods for handling time series. To go any further we need to look at the probability structure of the families of series that we intend to model in a little more detail. We will not be getting into very deep mathematics, but we will still have to look at a certain amount of formal structure.

There are some interesting theoretical problems about the distributions of possibly infinite sets of random variables which we shall ignore. We make a simplifying assumption, that is that the series we deal with are stationary or can be made stationary. By stationarity we mean that the series looks much the same over all time periods. More specifically the broad statistical properties are the same if we change the time origin. At a technical level we can have

- strong stationarity: this means that the joint distributions of any set of $X_{t_1}, X_{t_2}, \ldots, X_{t_k}, \ldots$ is unchanged if the times $t_1, t_2, \ldots$ are shifted by some amount s. This is both a very restrictive assumption and it is not very useful in practice
- weak stationarity: for weak stationarity rather than fix the entire distribution we only require the first two moments of the joint distributions to be the same for time shifts, that is $E[X_t X_s]$ is a function of $|t - s|$ alone.

We will use weak stationarity extensively and we can use the restrictions on the moments to define the mean of a weakly stationary series $\mu = E[X_t]$ and also the autocovariance sequence

$$\gamma(k) = E[(X_t - \mu)(X_{t+k} - \mu)] \qquad k = \ldots -1, 0, 1, 2, \ldots.$$

These are often scaled to give correlations

$$\rho(k) = \frac{\gamma(k)}{var(X_t)} \qquad k = \ldots -1, 0, 1, 2, \ldots.$$

The k values are often called the *lags*. It should be clear that the variance of the (stationary) series is $\gamma(0)$ and hence $\rho(k) = \gamma(k)/\gamma(0)$.

The autocorrelation can be thought of as the correlation between the original series and a copy displaced by k time points or k lags. They are, as we shall see, a measure of the series memory. Some simple properties of the autocorrelations are easily established

1. $\gamma(k) \le \gamma(0)$ or $|\rho(k)| \le 1$
2. $\rho(k) = \rho(-k)$
   Note this result holds for real series, for complex we need
   $\rho(k) = \overline{\rho(-k)}$
3. The $\rho(k)$ are a positive definite sequence – this is a means that any matrix whose $(ij)^{th}$ element is $\rho(|i - j|)$ has a non-negative determinant. This rather esoteric result will be useful later.

Despite our definitions it is clear that we have a problem. For most problems we have one realization $\{x_t\}$ from the series $\{X_t\}$. If we think in usual multivariate statistics terms we have a single point in n-space. *We have a sample of size 1!*

All is not lost for we can appeal to some rather esoteric theory to help. The ideas, if not the proofs, are not difficult, consider the mean $\mu = E[X_t]$. If this is constant over time it seems reasonable to estimate it using $\hat{\mu} = \frac{1}{N} \sum_{t=1}^{N} x_t$. If we have a stationary series then stationarity allows us to equate the averages at time t to the averages along the realization.

We can reduce our requirement to the following. Provided $\gamma(k) \to 0$ as $k \to \infty$ and that $\sum_{k=0}^{\infty} \gamma^2(k) \le \infty$ then

- $\hat{\mu} = \frac{1}{N} \sum_{t=1}^{N} x_t \to \mu$
- $\hat{\gamma}(k) = \frac{1}{N} \sum_{t=1}^{N-|k|} (x_t - \hat{\mu})(x_{t+|k|} - \hat{\mu}) \to \gamma(k)$

In passing we note that there are two forms of the estimate $\hat{\gamma}(k)$, the one given above and one which is unbiased with a divisor $N - |k|$ rather than N. When N is large there is little practical difference, we prefer the biased form for its simplicity and for other more esoteric reasons.

3.1 Determinism and predictability

Stationarity, in the second order sense is one of the central concepts of time series analysis, the other is predictability. This is again a fairly intuitive concept. Suppose we have $\{X_t\}$ then we consider the problem of predicting the series at time t+1 given the past of the series from time t into the past. We write the predictor of X_{t+k} made at time t, assuming knowledge of the past at t, by $\hat{X}_{t+k|t}$. Occasionally we will get lazy and write $\hat{X}_s$ when we mean $\hat{X}_{s|s-1}$. The squared error we make with a predictor $\hat{X}_{t+1|t}$ is

$$|e_{t+1}|^2 = |X_{t+1} - \hat{X}_{t+1|t}|^2 \tag{3.1}$$

Suppose we use the optimum linear predictor to make our predictions, then there are two quite distinct cases

1. $e_{t+1} = 0$ and our predictions are perfect. In this, rather uninteresting, case we say that the series is deterministic.
2. The more common position is that $e_{t+1} \neq 0$ and we have a non-deterministic series.

The idea of determinism is a central one and gives us a critical result proved in 1939 by Wold. We make no attempt at proof and we give the result below.

3.2 Wold's theorem

Suppose that a stationary series $\{X_t\}$ is non deterministic, then it has the representation

$$X_t = D_t + \epsilon_t + \sum_{j=1}^{\infty} g_j \epsilon_{t-j} \qquad (3.2)$$

Where $\{\epsilon_t\}$ is a zero mean white noise series and D_t is a deterministic series.

The part of the representation which is not perfectly predictable is the sum of noise terms. While this description of the series as a set of innovations ϵ_t is interesting in itself it is important as it gives a way of looking at the prediction process.

3.3 Prediction

We now consider the prediction of a non-deterministic series . Suppose we have a predictor of X_{t+k}, say $\hat{X}_{t+k|t}$

$$\hat{X}_{t+k|t} = a_1 X_t + a_2 X_{t-1} +, \ldots, + a_m X_{t-m}$$

We assume that this linear predictor, which may have an infinite number of terms is optimum in that the mean square error

$$\sigma_k^2 = E[e_k^2] = E\left[(X_{t+k} - \hat{X}_{t+k|t})^2 \right] \qquad (3.3)$$

is a minimum. Our aim is to find the form of this predictor. We point out two practical points here.

1. We are assuming a linear predictor.
2. We assume that the entire past from time t is available.

Since our predictor is made up of past values of the series it can be written as a sum of innovations, using Wold's theorem. Thus

$$\hat{X}_{t+k|t} = \sum_{j=0}^{\infty} b_j \epsilon_{t-j} \qquad (3.4)$$

We can therefore write the error as

$$\acute{e}_k = X_{t+k} - \hat{X}_{t+k|t} = \sum_{j=0}^{\infty} g_j \epsilon_{t+k-j} + \sum_{j=0}^{\infty} b_j \epsilon_{t-j}$$

This can be rewritten as

$$g_0 \epsilon_{t+k} + g_1 \epsilon_{t+k-1} + \cdots + g_{k-1} \epsilon_{t+1} - \sum_{j=0}^{\infty} (g_{j+k} - b_j) \epsilon_{t-j}$$

where of course $g_0 = 1$. If we now take expectations we have, since $E[\epsilon_r \epsilon_s] = \sigma^2$ when $r = s$ and is zero otherwise.

$$\sigma_k^2 = E[e_k^2] = \sum_{j=0}^{k-1} g_j^2 \sigma^2 + \sum_{j=0}^{\infty} (g_{j+k} - b_j)^2 \sigma^2$$

This will be a minimum when $g_{j+k} - b_j = 0$ in which case

1. The optimal predictor is

$$\hat{X}_{t+k|t} = g_k \epsilon_t + g_{k+1} \epsilon_{t-1} + g_{k+2} \epsilon_{t-2} + \cdots \qquad (3.5)$$

2. The error of prediction is

$$\epsilon_{t+k} + g_1 \epsilon_{t+k-1} + \cdots + g_{k-1} \epsilon_{t+1} \qquad (3.6)$$

3. The minimum mean square error of prediction is

$$\sigma_k^2 = \sigma^2 \sum_{j=0}^{k-1} g_j^2 \qquad (3.7)$$

It is worth bearing in mind that in the case where $k = 1$, that is one step ahead prediction, the error is ϵ_{t+1} and the mean square error is σ^2, where σ^2 is the innovation variance.

If we know something about the structure of our models then we can construct the optimum predictor.

3.4 Some examples of prediction

1. If we have a sequence of uncorrelated random variables, all having zero mean and common variance σ^2 then we would not expect to be able to predict the series. Here $X_t = \epsilon_t$ so we have a short innovations series! Any future value is $X_{t+k} = \epsilon_{t+k}$ but at time t we only know about current ϵ_t and past values. Thus our prediction is just zero.

2. Suppose we know that the generating structure of our model is

$$X_t = \epsilon_t + \theta\epsilon_{t-1}$$

where the $\{\epsilon_t\}$ series is zero mean while noise. Then if we are making predictions at time t, since

$$X_{t+k} = \epsilon_{t+k} + \theta\epsilon_{t+k-1}$$

our best predictor when $k > 1$ is $\hat{X}_{t+1|t} = 0$ from equation 3.5. When we wish to predict one step ahead then as

$$X_{t+1} = \epsilon_{t+1} + \theta\epsilon_t$$

our best predictor is $\theta\epsilon_t$. This is not much use as it stands since the observed values are X_t, but we know that **the innovations are the one step ahead prediction errors** so

$$\epsilon_t = X_t - \hat{X}_{t|t-1}$$

giving

$$\hat{X}_{t+1|t} = \theta(X_t - \hat{X}_{t|t-1})$$

We can easily see

$$\hat{X}_{t+1|t} = \theta X_t - \theta^2 X_{t-1} + \theta^3 X_{t-2} - \cdots - \theta^{m+1}(X_{t-m} - \hat{X}_{t-m|t-m-1})$$

and for $|\theta| \leq 1$ we can ignore the hatted term when we take m sufficiently large.

3. Suppose now

$$X_t = \phi X_{t-1} + \epsilon_t$$

where the ϵ_t is a zero mean white noise process and $E[X_t\epsilon_{t+s}] = 0$ when $s > 0$. If we manipulate the equation we have

$$X_t = \epsilon_t + \phi\epsilon_{t-1} + \phi^2\epsilon_{t-2} + \phi^3\epsilon_{t-3} + \cdots + \phi^t\epsilon_0 + \phi^t X_0$$

The k step ahead predictor is thus

$$\phi^k\epsilon_t + \phi^{k+1}\epsilon_{t-1} + \cdots$$

which after some manipulation turns out to be $\phi^k X_t$. *As we shall see there are simpler approaches.*

3.5 The ARMA models

Models for time series based on difference equations are widely used and are associated with the pioneering work of Box and Jenkins (1970) who introduced many central ideas. We begin by looking at the components of these models and in particular with the moving average model.

3.5.1 The moving average model

As we have seen any nondeterministic model can be written as a, possibly infinite, sum of innovations. Thus considering a finite sum as a model for a time series is quite natural. A model of the form

$$X_t = \epsilon_t + \theta_1\epsilon_{t-1} + \theta_2\epsilon_{t-2} + \cdots + \theta_q\epsilon_{t-q} \tag{3.8}$$

where $\{\epsilon_t\}$ is zero mean white noise is called a moving average process of order q, or more commonly an MA(q) process. We can of course incorporate a mean to give

$$X_t = \mu + \epsilon_t + \theta_1\epsilon_{t-1} + \theta_2\epsilon_{t-2} + \cdots + \theta_q\epsilon_{t-q}$$

if we feel that this is required, however we usually work with series from which the mean has been removed. For simplicity in manipulation we often use as a shorthand the notation

$$X_t = \theta(B)\epsilon_t \tag{3.9}$$

where $\theta(B)$ is a polynomial of order q in the backshift operator B that is

$$\theta(B) = 1 + \theta_1 B + \theta_2 B^2 + \cdots + \theta_q B^q$$

We can let the first term be one as we have the series variance as an extra parameter.

The autocovariances of these series take the form

$$\gamma(k) = E[X_t X_{t+k}] = E\left[\sum_{j=0}^{q}\theta_j\epsilon_{t-j}\sum_{j=0}^{q}\theta_j\epsilon_{t+k-j}\right] = E\sum_{i=0}^{q}\sum_{j=0}^{q}\theta_i\epsilon_{t-i}\theta_j\epsilon_{t+k-j} \tag{3.10}$$

but $E[\epsilon_{t-i}\epsilon_{t+k-j}] = 0$ unless $t - i = t + k - j$ or $j = i + k$. This means that we have

$$\gamma(k) = E[X_t X_{t+k}] = \sigma^2 \sum_{j=0}^{q}\theta_i\theta_{i+k} \tag{3.11}$$

when $k \leq q$ and zero otherwise. Note that not only have we set θ_0 to one but also assume that $\theta_j = 0$ when $j > q$. The variance is therefore

$$var[X_t] = \sigma^2 \sum_{j=0}^{q}\theta_i^2 \tag{3.12}$$

The two interesting points that follow immediately are

1. Once the value of the the lag exceeds q the autocovariances are zero, that is $\gamma(j) = 0$ when $j > q$
2. The moving average process is stationary.

Moving average examples

The simplest possible model is the MA(0) which is just an uncorrelated sequence of random variates, known as a white noise process. It has zero autocorrelations, apart of course from the zero order autocorrelation! A more instructive example is the moving average process of order two, the MA(2) process

$$X_t = \epsilon_t + \theta_1 \epsilon_{t-1} + \theta_2 \epsilon_{t-2}$$

In this case

$$\gamma(0) = \sigma^2(1 + \theta_1^2 + \theta_2^2)$$
$$\gamma(1) = \sigma^2(\theta_1 + \theta_1\theta_2)$$
$$\gamma(2) = \sigma^2\theta_2$$

and

$$\gamma(k) = 0 \text{ for } |k| > 2$$

We often use the backshift operator B when writing models as it can simplify our notation. The general MA(q) model is

$$X_t = (1 + \theta_1 B + \theta_2 B^2 + \theta_3 B^3 + \cdots + \theta_q B^q)\epsilon_t = \theta(B)\epsilon_t$$

where $\theta(B)$ is a polynomial in B. Thus for our simple case above

$$X_t = \epsilon_t + \theta_1 \epsilon_{t-1} + \theta_2 \epsilon_{t-2} = (1 + \theta_1 B + \theta_2 B^2)\epsilon_t$$

the polynomial $\theta(B)$ is

$$\theta(B) = 1 + \theta_1 B + \theta_2 B^2$$

*To illustrate the value of these polynomials we quote a result: the MA(q) process can be forecast provided the roots of the polynomial $\theta(B)$ **lie outside the unit circle**. This is known as the invertibility condition.*

We shall try to show the relevance of this condition by considering MA(2) model above. We have

$$X_{t+1} = \epsilon_{t+1} + \theta_1 \epsilon_t + \theta_2 \epsilon_{t-1}$$

so if we wish to forecast the series at time t+1 given time t we have using the prediction equation 3.5 and $\epsilon_t = X_t - X_{t|t-1}$

$$\hat{X}_{t+1|t} = \theta_1(X_t - \hat{X}_{t|t-1}) + \theta_2(X_{t-1} - \hat{X}_{t-1|t-2})$$

To get forecasts we need to substitute for $\hat{X}_{t|t-1}$ in terms of $\hat{X}_{t-1|t-2}$ and $\hat{X}_{t-1|t-2}$ in terms of $\hat{X}_{t-2|t-3}$ and so on. If we can eventually forget the far past we can make successful predictors.

The invertibility condition is just a way of saying that prediction is possible. If we think in terms of $\theta(B)$ then consider the roots of the polynomial, say α_1 and α_2. This means

$$\theta(B) = (\alpha_1 - B)(\alpha_2 - B)$$

We can write

$$X_t = (\alpha_1 - B)(\alpha_2 - B)\epsilon_t$$

Now in a purely formal way we could consider writing this as

$$(\alpha_1 - B)^{-1}(\alpha_2 - B)^{-1} X_t = \epsilon_t$$

To expand $(\alpha_1 - B)^{-1}(\alpha_2 - B)^{-1}$ we need to write it as a partial fraction and then expand the components. We find that if we do a formal expansion we need $1/\alpha_1 < 1$ and $1/\alpha_2 < 1$ for a finite or bounded value. If we can expand in this way then the series is written in terms of its past and so we may forecast. The condition is just the invertibility one! This rather mathematical argument is complex but does give us an insight into the general case. The reader can of course just take us at our word!

3.5.2 The autoregressive model

The moving average model is rather indirect and a more obvious model is the autoregressive model where the value of the series at time t depends on previous values. More specifically the autoregressive model of order p AR(p), is

$$X_t = \phi_1 X_{t-1} + \phi_2 X_{t-2} + \phi_3 X_{t-3} + \cdots + \phi_p X_{t-p} + \epsilon_t \qquad (3.13)$$

where the $\{\epsilon_t\}$ series is zero mean white noise and $E[X_t \epsilon_{t+s}] = 0$ when $s > 0$. Here we have a very direct model with obvious roots in regression models. 'again this is often written in terms of a polynomial in the backshift operator B. The form is

$$\phi(B)X_t = \epsilon_t \qquad (3.14)$$

The polynomial $\phi(B)$, of order p is sometimes called the AR or autoregressive polynomial and to fit with our definition we have

$$\phi(B) = 1 - \phi_1 B - \phi_2 B^2 - \phi_3 B^3 - \cdots - \phi_p B^p$$

The obvious question to ask is, are AR models stationary? The answer is that they are not unless the AR polynomial

$$\phi(B) = 1 - \phi_1 B - \phi_2 B^2 - \cdots - \phi_p B^p$$

satisfies certain conditions. We quote the result without proof: *The AR(p) model is stationary if and only if the roots of the polynomial $\phi(B)$ lie outside the unit circle.* So for a simple AR(1) model

$$X_t = \phi X_{t-1} + \epsilon_t$$

this is stationary provided $|\phi| < 1$. The reason is clear in this case since

$$X_t = \phi X_{t-1} + \epsilon_t = \phi(\phi X_{t-2} + \epsilon_{t-1}) + \epsilon_t = \epsilon_t + \phi\epsilon_{t-1} + \cdots + \phi^k\epsilon_{t-k} + \cdots$$

and the series above will have an infinite variance if $|\phi|$ is one or more. But $\phi(B) = 1 - \phi B$ which has a root $1/\phi$ and this is greater than one if $|\phi| < 1$

Example 3.1. We have already looked at the AR(1) model when we discussed forecasting

$$X_t = \phi X_{t-1} + \epsilon_t \tag{3.15}$$

To get the autocorrelations we consider the model equation multiplied by X_{t-k}. If we take expectations we have

$$E[X_t X_{t-k}] = E[\phi X_{t-1} X_{t-k} + \epsilon_t X_{t-1}]$$

since we assume that the future innovations are independent of the current observations

$$\gamma(k) = \phi\gamma(k-1)$$

In consequence $\gamma(|k|) = \phi^{|k|}\gamma(0)$. To get at the variance $\gamma(0)$ we need to be a bit more subtle. If we multiply equation 3.15 by ϵ_t we have

$$X_t\epsilon_t = X_{t-1}\epsilon_t + \epsilon_t\epsilon_t$$

which given, on taking expectations, $E[X_t\epsilon_t] = \sigma^2$. If we now multiply equation 3.15 by X_t we get after taking expectations

$$var[X_t] = \sigma^2/(1-\phi^2)$$

3.5.3 The Yule–Walker equations

If we generalize the argument we have just followed for the first order, AR(1), process we obtain for the AR(p) process a system of equations for the autocovariances and autocorrelations. This set of equations is known as the Yule-Walker equations. The procedure is simple, take the equation 3.13 and multiply by X_{t-k}.

$$E[X_t X_{t-k}] = E[\phi_1 X_{t-1} X_{t-k} + \phi_2 X_{t-2} X_{t-k} + \cdots + \phi_p X_{t-p} X_{t-k} + \epsilon_t X_{t-k}]$$

Taking expectations gives

$$\gamma(k) = \phi_1\gamma(k-1) + \phi_2\gamma(k-2) + \phi_3\gamma(k-3) + \cdots + \phi_p\gamma(k-p)$$

So setting $k = 1, 2, \ldots$, we have

$$\gamma(1) = \phi_1\gamma(0) + \phi_2\gamma(1) + \phi_3\gamma(2) + \cdots + \phi_p\gamma(p-1) \tag{3.16}$$
$$\gamma(2) = \phi_1\gamma(1) + \phi_2\gamma(0) + \phi_3\gamma(1) + \cdots + \phi_p\gamma(p-2)$$

$$\cdots\cdots\cdots\cdots$$

$$\cdots\cdots\cdots\cdots$$

$$\gamma(p) = \phi_1\gamma(p-1) + \phi_2\gamma(p-2) + \phi_2\gamma(p-3) + \cdots + \phi_p\gamma(0)$$

The equations are rather neater in terms of the autocorrelations.

$$\rho(k) = \phi_1 \rho(k-1) + \phi_2 \rho(k-2) + \phi_3 \rho(k-3) + \cdots + \phi_p \rho(k-p)$$

This gives a set of equations

$$\rho(1) = \phi_1 + \phi_2 \rho(1) + \phi_3 \rho(2) + \cdots + \phi_p \rho(p-1) \qquad (3.17)$$
$$\rho(2) = \phi_1 \rho(1) + \phi_2 + \phi_3 \rho(1) + \cdots + \phi_p \rho(p-2)$$

$$\cdots\cdots\cdots\cdots\cdots$$

$$\cdots\cdots\cdots\cdots\cdots$$

$$\rho(p) = \phi_1 \rho(p-1) + \phi_2 \rho(p-2) + \cdots + \phi_p$$

In matrix form they become

$$
\begin{pmatrix} \rho(1) \\ \rho(2) \\ \cdots \\ \rho(p) \end{pmatrix}
=
\begin{pmatrix}
1 & \rho(1) & \rho(2) & \cdots & \rho(p-1) \\
\rho(1) & 1 & \rho(1) & \cdots & \rho(p-2) \\
& & \cdots & & \\
\rho(p-1) & \rho(p-2) & \rho(p-3) & \cdots & 1
\end{pmatrix}
\begin{pmatrix} \phi_1 \\ \phi_2 \\ \cdots \\ \phi_p \end{pmatrix}
$$

say

$$\mathbf{r_p} = \mathbf{R_p}\phi \qquad (3.18)$$

You will notice that these equations are just the left hand side of the difference equations. We can solve these equations but if we are to consider the stationarity and prediction of these series we need to look at the difference equation and this is simpler if we consider the AR(p) model as

$$\phi(B)X_t = \epsilon_t$$

where $\phi(B)$ is a polynomial in the backshift operator B.

3.6   Difference equations and stationarity

Some of the ideas we need are most easily obtained by looking at difference equations. For an equation of the form

$$(1 - \phi B)x_t = 0$$

we can easily verify that a solution is ϕ^t. It is then possible to solve a second order equation, for $(1 - \phi_1 B - \phi_2 B^2)X_t = 0$ can be written as $(1 - \alpha_1 B)(1 - \alpha_2 B)X_t = 0$ with a solution $A\alpha_1^t + B\alpha_2^t$ where A and B are constants. The extension to polynomials of higher order in B is direct.

When the right hand side is non-zero then we shall just assume that we can expand the inverse. So if $(1 - \phi B)X_t = \epsilon_t$ we just manipulate the polynomial in B to give

$$X_t = (1 - \phi B)^{-1}\epsilon_t = (1 + \phi B + \phi^2 B^2 + \phi^3 B^3 + \cdots)\epsilon_t$$

At this point we must remember that the solutions of $(1 - \phi B)X_t = 0$ will also be solutions of our equation so we add these to give our solution

$$X_t = \phi^t + (1 + \phi B + \phi^2 B^2 + \phi^3 B^3 + \cdots)\epsilon_t$$

Of course we assume that the solution exist or that the series we have used in the expansion does converge. The criterion in this case is $|\phi| < 1$.

This gives us direct way of looking at the stationarity of an autoregressive process. Clearly any AR model can be written in terms of a polynomial in B which we can then factorize,

$$\phi(B)X_t = \prod(\alpha_j - B)X_t = \epsilon_t$$

The series will be stationary provided the α_j exceed one in modulus. This is equivalent to saying that the roots of $\Phi(z)$ **lie outside the unit circle**.

3.6.1 Examples of AR models, MA models using difference equations

Consider the ARMA model

$$x_t - \frac{2}{3}x_{t-1} + \frac{1}{12}x_{t-2} = \epsilon_t$$

We can write this as

$$\left(1 - \frac{2}{3}B + \frac{1}{12}B^2\right)x_t = \left(1 - \frac{1}{2}B\right)\left(1 - \frac{1}{6}B\right)x_t = \epsilon_t$$

Since the multipliers of the B's are less that 1 we know that the series is stationary. If we solve $\Phi(z) = 1 - \frac{2}{3}z + \frac{1}{12}z^2 = 0$ the roots are $1/\frac{1}{2}$ and $1/\frac{1}{6}$ both of which lie outside the unit circle implying stationarity. The Yule–Walker equations are just

$$\left(1 - \frac{2}{3}B + \frac{1}{12}B^2\right)\gamma(k) = 0$$

or in terms of the autocorrelations

$$\left(1 - \frac{2}{3}B + \frac{1}{12}B^2\right)\rho(k) = 0$$

Thus the solutions are

$$\rho(k) = A\left(\frac{1}{2}\right)^k + B\left(\frac{1}{6}\right)^k$$

We can then use boundary conditions to solve.

Since $\rho(0) = 1$ we know A+B=1 and $\rho(1) = \rho(-1)$ thus

$$A\left(\frac{1}{2}\right) + B\left(\frac{1}{6}\right) = A\bigg/\left(\frac{1}{2}\right) + B\bigg/\left(\frac{1}{6}\right)$$

Solving gives $A = 35/26$ and $B = -9/26$ which allows us to compute the autocorrelations. Whether this is any simpler than solving the Yule–Walker equations is debatable, what is worth noting is that the autocorrelations are made up of term which decay exponentially.

This example was a simple one, we should not forget the complex solutions. Consider

$$\left(1 - B + \frac{13}{16}B^2\right)x_t = \epsilon_t$$

The solutions to $1 - z + \frac{13}{16}z^2 = 0$ are $1/2 + i3/4$ and $1/2 - i3/4$. We can rewrite them as $\frac{\sqrt{13}}{4}exp(i\theta)$ and $\frac{\sqrt{13}}{4}exp(-i\theta)$ where $\theta = tan^{-1}\left(\frac{3}{2}\right)$. Clearly the roots lie in the unit circle so the series is stationary. The form of the autocorrelations is

$$\rho(k) = A\left(\frac{\sqrt{13}}{4}\right)^k exp(i\theta k) + B\left(\frac{\sqrt{13}}{4}\right)^k exp(-i\theta k)$$

This means that the correlations will behave like a damped sinusoid. If we use the boundary condition that $\rho(0) = 1$ we get $\rho(k) = \left(\frac{\sqrt{13}}{4}\right)^k cos(\theta k)$

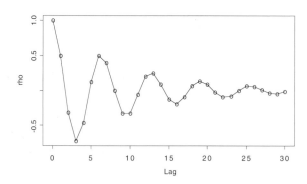

Fig. 3.1 *Autocorrelation with damped oscillations*

This is an oscillation in the autocorrelations, shown in Figure 3.1 and indicates that there must be a similar oscillation in the series itself and we must beware assuming that it is a cyclic effect.

We discussed invertibility for MA processes earlier in this section.

$$X_t = \epsilon_t + \theta\epsilon_{t-1}$$

is just

$$X_t = (1 + \theta B)\epsilon_t$$

If we use operators in a natural way we have

$$(1 + \theta B)^{-1}X_t = \epsilon_t$$

so

$$(1 + \theta B + \theta^2 B^2 + \theta^3 B^3 + \cdots)X_t = \epsilon_t$$

and the left hand side will only be sensible when $|\theta| < 1|$.

3.7 The autoregressive moving average model

Both AR and MA model have been used in time series analysis but it seems natural to put them together to get the rather more general ARMA autoregressive-moving average model. The ARMA(p,q) model is just

$$X_t = \phi_1 X_{t-1} + \phi_2 X_{t-2} + \cdots + \phi_p X_{t-p} + \epsilon_t + \theta_1\epsilon_{t-1} + \theta_2\epsilon_{t-2} + \cdots + \theta_q\epsilon_{t-q} \tag{3.19}$$

or in polynomial form

$$\phi(B)X_t = \theta(B)\epsilon_t \tag{3.20}$$

We have $\{\epsilon_t\}$, uncorrelated zero mean series and we assume that $E[X_t\epsilon_{t+s}] = 0$ when $s > 0$, *that is future noise terms are uncorrelated with past terms of the series*. This is an entirely reasonable assumption.

We can deduce some properties of this ARMA(p,q) model from our earlier discussion.

- The series is stationary if and only if the roots of $\phi(z)$ lie outside the unit circle.
- The series is invertible if and only if the roots of $\theta(z)$ lie outside the unit circle.
- The series satisfies the same Yule–Walker equations as an AR(p) model for sufficiently high lags.
- We know that for an AR series the predictions satisfy the same difference equation as the AR series. This is also true of the ARMA model for predictions sufficiently far into the future.

Now we know we can convert an ARMA model into a pure MA or a pure AR to make predictions. A simpler way is to look at the components. Take the following ARMA(2,1) model.

$$X_t = \phi_1 X_{t-1} + \phi_2 X_{t-2} + \epsilon_t + \theta\epsilon_{t-1}$$

If we are at time t and want to forecast values at t+k we can write

$$X_{t+k} = \phi_1 X_{t+k-1} + \phi_2 X_{t+k-2} + \epsilon_t + \theta \epsilon_{t+k-1}$$

now at time t X_{t+k} is the forecast $\hat{X}_{t+k|t}$ since it is a future value. In the same way ϵ_{t+k} is zero for the innovations are not predictable except as a zero mean. Table 3.1 gives these for $s > 0$

Table 3.1

Value	X_{t+s}	X_{t-s}	ϵ_{t+s}	ϵ_{t-s}
Forecast at time t	$\hat{X}_{t+s|t}$	X_{t-s}	0	$X_{t-s} - \hat{X}_{t-s|t-s-1}$

3.7.1 Prediction intervals

Given the model the construction of forecasts is reasonably simple but one would also like some idea of the prediction errors. An simple approximation is possible as follows. Suppose our model is written in Moving average (innovations form)

$$X_t = \epsilon_t + \psi_1 \epsilon_{t-1} + \psi_2 \epsilon_{t-2} + \psi_3 \epsilon_{t-3} + \cdots$$

then we know that the minimum mean square error of prediction is $\sigma^2 = var(\epsilon_t)$ for a one step ahead forecast and $\sigma^2(1 + \sum_{j=1}^{k-1} \psi_j^2)$ for k steps ahead. If we assume normality for the forecast function we have a set of confidence intervals for our forecast, these so called prediction intervals are for X_{t+k} at time t

$$\hat{X}_{t+k|t} \pm z_{\alpha/2} \sqrt{\left[1 + \sum_{j=1}^{k-1} \psi_j^2 \sigma^2 \right]}$$

Unfortunately experience shows that the resulting confidence intervals are too narrow. They give some idea but should be used only as a guide.

The exponential smoothing model

As an example we can show that the exponential smoothing model is optimum for a model of the form

$$X_t = X_{t-1} + \epsilon_t + \theta \epsilon_{t-1}$$

Suppose we forecast this model one step ahead, the model formula is $X_{t+1} = X_t + \epsilon_{t+1} + \theta \epsilon_t$ so if we are standing at time t we have

$$\hat{X}_{t+1|t} = X_t + \theta \epsilon_t = X_t + \theta(X_t - \hat{X}_{t|t-1})$$

if we collect terms and write $\alpha = 1 - \theta$ we have the exponential smoothing form from Chapter 2. This implies that simple exponential smoothing is in fact a special case of the ARMA type model. As we shall see this is called an ARIMA(0,1,1) model.

3.7.2 Yule–Walker equations revisited

Given an ARMA model we can construct the equations for the autocorrelations. Suppose we have

$$X_t = \phi_1 X_{t-1} + \cdots + \phi_p X_{t-p} + \epsilon_t + \theta_1 \epsilon_{t-1} + \cdots + \theta_q \epsilon_{t-q}$$

Now we know that values of X are uncorrelated with future noise terms so we can multiple the equation by X_{t-k} and take expectations we have

$$X_{t-k} X_t = \phi_1 X_{t-k} X_{t-1} + \cdots + X_{t-k} \phi_p X_{t-p} + X_{t-k} \epsilon_t$$
$$+ \theta_1 X_{t-k} \epsilon_{t-1} + \cdots + \theta_q X_{t-k} \epsilon_{t-q}$$

Provided $k > q$ we have

$$\gamma(k) = \phi_1 \gamma(k-1) + \cdots + \phi_p \gamma(k-p) = 0 \text{ for } k = q+1, q+2 \ldots$$

This is just the set of equations for the AR part of the model. Of course there is a difference for the initial equations. It is simplest to see with an example. Take

$$X_t = \phi_1 X_{t-1} \epsilon_t + \theta_1 \epsilon_{t-1}$$

If we now multiply by ϵ_t we have on taking expectations

$$X_t \epsilon_t = \phi_1 \epsilon_t X_{t-1} + \epsilon_t \epsilon_t + \theta_1 \epsilon_t \epsilon_{t-1}$$

on taking expectations

$$E[X_t \epsilon_t] = \sigma^2$$

If we now multiply by ϵ_{t-1}

$$X_t \epsilon_{t-1} = \phi_1 \epsilon_{t-1} X_{t-1} + \epsilon_{t-1} \epsilon_t + \theta_1 \epsilon_{t-1} \epsilon_{t-1}$$

giving

$$E[X_t \epsilon_{t-1}] = \phi_1 E[\epsilon_t X_{t-1}] + \theta_1 \sigma^2$$

or

$$E[X_t \epsilon_{t-1}] = (\phi_1 + \theta_1) \sigma^2$$

This enables us to find the first few equations in the Yule–Walker set. It is worth noticing that for large lags we have the Yule–Walker equations for a pure AR process implying that for log time intervals the process can be regarded as AR.

3.7.3 Simulation

One of the great advantages of developing the ARMA model is the ability to simulate series. In any programming language once one has a means of generating random numbers one can very simply generate ARMA models using the difference equations. We have provided function AR,ma and arma which will generate series using the appropriate model. It is often useful to be able to try ideas on 'known' series or to see how a series actually looks. Thus we looked at the ar(2) model

$$\left(1 - B + \frac{13}{16}B^2\right)x_t = \epsilon_t$$

In R (doubtless the reader can also write simple code in an alternative) we can generate 100 observations from this model and plot them and the autocovariances using

```
x<-ar(c(2/3,-1/12),100)
tsplot(x)
acfplot(x,25)
```

The resulting series will look rather like one generated by 3.2. Not all that cyclic but with an indication of a cycle in the autocorrelations.

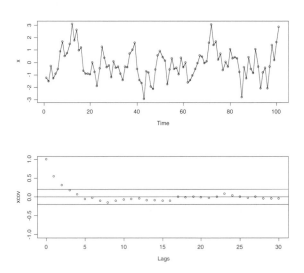

Fig. 3.2 *A simulation using an AR(2) model*

You might like to generate some optimum models for exponential smoothing and then try them out!

3.7.4 The ARIMA model

The reader will have noticed a gap in our narrative. We have theoretical models for series which must be stationary. In the real world, however, the series we observe are very often not stationary, they have trends. The trick is to extend the ARMA class to enable us to come up with a model for a non-stationary series.

Suppose we have a series $\{X_t\}$ with a linear trend. If we construct the differences $X_t - X_{t-1}$, that is $(1 - B)X_t$ it is easy to see that the difference series no longer has a trend. Indeed if we have a series with a trend following a polynomial of order k then we can show that applying $(1 - B)^k$ removes the trend. This gives us clues to a non-stationary model for differencing which may give a stationary and ARMA form. However series with deterministic trends are not the only type of non-stationary series. Suppose we have the random walk model

$$(1 - B)X_t = \epsilon_t \tag{3.21}$$

that is a series which when differenced is white noise. If we manipulated 3.21 we get

$$X_t = \epsilon_t + \epsilon_t+, \dots, +\epsilon_1 + X_0$$

Then the mean is constant but the variance of the series depends on the time since $var[X_t] = t\sigma^2 + var[X_0]$, indeed if we modify 3.21 to include a mean

$$(1 - B)X_t = \mu + \epsilon_t$$

then we get a stochastic trending mean as well as a trending variance. The model

$$\phi(B)(1 - B)^d X_t = \theta(B)\epsilon_t$$

is known as a ARIMA(p,d,q) model (autoregressive integrated moving average) and is used to model non-stationary series as we shall see. It is instructive to look at a simple form.

$$(1 - B)X_t = \epsilon_t$$

may be written as

$$X_t = (1 - B)^{-1}\epsilon_t = (1 + B + B^2 + B^3 + \cdots)\epsilon_t$$

The variance here is not bounded, if we start from time t we have a build up in variance and we conclude that the series cannot be stationary. An artificial example is shown in Figure 3.3

If we consider the Yule–Walker equations we find that for large lags the autocorrelations satisfy the difference equation

$$(1 - B)^d \phi(B)\rho(k) = 0$$

The zeros on the right hand side will include one value which has modulus one. That means that in general the autocorrelations do not decay exponentially as in the stationary case. This is pretty obvious at an intuitive level.

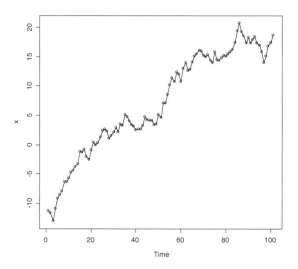

Fig. 3.3 *A simulation of a random walk model*

Informally we can think of the ARIMA model as one where the series is differenced d times to a stationary form and then an ARMA model fitted to the stationary part.

In Chapter 2 we discussed exponential smoothing and we are now in a position to show that the Holt Winters approach is in fact a special case of an ARIMA(0,2,2) model. Recall we forecast X_{t+1} by $M_t + T_t$ where

$$M_t = AX_t + (1 - A)(M_{t-1} + T_{t-1}) \text{ and } T_t = C(M_t + M_{t-1}) + (1 - C)T_{t-1}$$

and if the model is correct then the one step ahead errors are $e_{t+1} = X_{t+1} - M_t - T_t$. If we manipulate the recursions we have

$$M_t - M_{t-1} = T_{t-1} + A(X_t - M_{t-1} - T_{t-1}) = T_{t-1} + Ae_t$$

In the same way

$$T_t - T_{t-1} = C(M_t - M_{t-1}) - CT_{t-1} = ACe_t$$

Now for the correct model $e_t = X_t - M_{t-1} - T_{t-1}$ so if we difference this equation twice

$$(1 - B)^2 e_t = (1 - B)^2 X_t - (1 - B)^2 M_{t-1} - (1 - B)^2 T_{t-1}$$

$$= (1 - B)^2 X_t - AC(1 - B)e_t + (1 - B)^2 Ae_t$$

Tiding up gives the result that we need.

3.8 Seasonal effects

You will have noticed we have made no comments about seasonal effects when discussing the AR and ARIMA models. What should one do in this case? The solution is obvious if not simple – extend the ARIMA model to the seasonal case.

Suppose we have a seasonal series with a seasonal period of s. Then the values of X_t and X_{t-s} are related so it may be worth considering their difference $X_t - X_{t-s}$ which may not have a seasonal effect. Thus, applying the 'seasonal difference' operator $1 - B^s$ a few times may remove seasonal effects giving a model

$$(1 - B^s)^D (1 - B)^d \phi(B) X_t = \theta(B)\epsilon_t$$

In practice this is never quite enough and it makes sense to add terms which related the seasonal effects. The simplest way to do this is to incorporate a seasonal ARMA effect to give the seasonal model

$$(1 - B^s)^D (1 - B)^d \phi(B) \Phi(B^s) X_t = \theta(B)\Theta(B^s)\epsilon_t \qquad (3.22)$$

where $\Phi(B^s)$ and $\Theta(B^s)$ are polynomials in B^s. Thus, some examples might be

$$(1 - B^4)(1 - 0.3B) X_t = \epsilon_t$$

or

$$(1 - B^{12})(1 - B)(1 - 0.4B + 0.7B^2) X_t = (1 - 0.4B^{12})\epsilon t$$

Once we have established the model there is little that is novel except perhaps for the complexity. We can of course just multiply out the polynomials and we are left with an ARIMA model. We will examine details of such models when we look at the fitting of seasonal models. In passing we will just point out to the reader who feels that we are being a little cavalier that the solutions of $(1 - B)^4 x_t = 0$ are a linear combination of the four roots of unity so the solution will have periodic effects.

We have spent some time looking at ARMA and other models in a purely theoretical way. We have made various assertions that these provide good models for real data series. Now is the time to put our claims to the test. In the next chapter we consider an alternative form of model. We can then look at some real data series. The reader who wishes to keep with the ARIMA theme can skip the next chapter and move on to where we deal with the problems of identifying and estimating the parameters of models for some real series series.

3.9 Exercises

1. Suppose that ϵ_t is zero mean white noise which of the following are stationary to second order?

    (a) $X_t = (-1)^t \epsilon_t$

(b) $X_t = \epsilon_t \cos(\omega t)$

(c) $X_t = \epsilon_t \epsilon_{t-1}$

(d) $X_t = A + Bt + \epsilon_{t-1}$

(e) $X_t = exp(i\omega t)\epsilon_t$

2. Suppose we define as series as

$$P[X_t = -1] = \frac{1}{2} \text{ and } P[X_t = 1] = \frac{1}{2}$$

If the successive terms of the series are independent what are the moments of $\{X_t\}$? Is the process stationary?

3. if we define a series as

$$X_t = \sum_{j=1}^{t} \epsilon_t$$

show that the series is not stationary.

4. Suppose

$$X_t = -1.6 + \epsilon_t - 0.6\epsilon_{t-1} + 0.3\epsilon_{t-2}$$

Is the process stationary? If $X_{93} = 3.8$ would you expect X_{94} to exceed this value?

5. Forecast the series given in question 4, 1, 2 and 12 steps ahead. What is the prediction error in each case?

6. Forecast

$$X_t = -1.6 + X_{t-2} + \epsilon_t - 0.6\epsilon_{t-1} + 0.3\epsilon_{t-2}$$

1, 2 and 12 steps ahead.

7. Generate a realization of length 200 from $X_t = 0.8X_{t-1} + \epsilon_t$. Does this look smoother that the white noise series ϵ_t? Compute the correlations and compare them with the expected values.

4
The state space approach

The difference equations of the ARIMA type have been very successful in time series and are widely used for modelling. There are however alternatives and we turn our attention to one of these, the *state space model*. These were proposed by several authors' times series, e.g. West and Harrison (1989), Harrison and Stevens (1976) and Harvey (1981) but have their roots in control engineering.

We suppose that out observed series $\{X_t\}$ can be written in terms of unobserved state variables, say a vector $\{\alpha_t\}$. The explicit form is

$$\mathbf{X}_t = \mathbf{H}\alpha_t + \epsilon_t \tag{4.1}$$

where $\mathbf{H}$ is a known matrix and ϵ_t is as usual white noise. To make life simpler we also assume that the state variables satisfy

$$\alpha_t = \phi\alpha_{t-1} + \mathbf{K}\eta_t \tag{4.2}$$

where ϕ is the transition matrix, and η_t is a noise vector, uncorrelated with ϵ_t

This rather curious set of equations can be justified in terms of conditional expectations in a multivariate normal distribution. However, the main reason for our interest is in the set of updating equations, the so called *Kalman recursions*. Before looking at these we consider some simple examples.

The simple random walk

Suppose we take the both H, K and transition matrix ϕ to be scalars taking the value 1 then the equations are just

$$X_t = \alpha_t + \epsilon_t$$

and

$$\alpha_t = \alpha_{t-1} + \eta_t$$

Our series X_t is generated by a random mean level with added noise.

We note that $(1 - B)\alpha_t = \eta_t$ and as

$$(1 - B)X_t = (1 - B)\alpha_t + (1 - B)\epsilon_t \tag{4.3}$$

this is equivalent to

$$(1 - B)X_t = \eta_t + (1 - B)\epsilon_t \tag{4.4}$$

The right hand side of this equation is $\eta_t + \epsilon_t - \epsilon_{t-1}$ which we can regard as two white noise components at time t and at time t-1. They both have, by definition, zero mean but have differing variances. We can write them as v_t and θv_{t-1}. Then

$$var[v_t] = var[\eta_t + \epsilon_t] = \sigma_\eta^2 + \sigma_\epsilon^2 \tag{4.5}$$

while

$$var[\theta v_{t-1}] = \sigma_\epsilon^2 \tag{4.6}$$

giving us the parameter

$$\theta = \sqrt{\frac{\sigma_\epsilon^2}{\sigma_\eta^2 + \sigma_\epsilon^2}} \tag{4.7}$$

This state space model is thus equivalent to an ARIMA(0,1,1). As we can see the construction is rather different and the state space formulation in terms of state variable gives a different flavour to our processes.

If we had discounted the state variables say $\alpha_t = \phi\alpha_{t-1} + \eta_t$ for some scalar $\phi \leq 1$ then we would have a model equivalent to an ARIMA(1,0,1) model.

Perhaps a more interesting case is the random walk but with a drifting mean. If we write the state variable α_t as a mean μ_t we have as before

$$x_t = \mu_t + \epsilon_t \tag{4.8}$$

but now we include a slope term β_t so that the mean drifts, that is

$$\mu_t = \mu_t + \beta_{t-1} + \eta_t \tag{4.9}$$

where

$$\beta_t = \beta_{t-1} + \zeta_t \tag{4.10}$$

In matrix terms we now have a vector of state variables $\alpha_t' = (\mu_t, \beta_t)$ with the recursion

$$\alpha_t = \begin{pmatrix} \mu_t \\ \beta_t \end{pmatrix} = \begin{pmatrix} 1 & 1 \\ 0 & 1 \end{pmatrix} \begin{pmatrix} \mu_{t-1} \\ \beta_{t-1} \end{pmatrix} + \begin{pmatrix} \eta_t \\ \zeta_t \end{pmatrix} \tag{4.11}$$

Again ϵ_t, η_t and ζ_t are noise terms.

Many authors, notably Harvey (1981) have used the state space equations explicitly giving rise to what are known as *structural equation models*. These are in reality a subset of the ARIMA family but are valuable in modelling terms as they give an alternative approach via state variables. This alternative view can be very useful.

The reverse is also true: we can rewrite ARMA models in state space form. While this is at first sight of rather limited interest it has one important implication, we can use the state space form to construct the likelihood function. This gives us a relatively simple and effective method of computing the likelihood. To see how this works we digress into some theory. This is to give us some insight into our main estimation methods.

With some work we can show that any ARMA(p,q) model can be written in state space form as

$$x_t = [1, 0, 0, \ldots, 0, 0]\alpha_t \tag{4.12}$$

$$\alpha_t = \begin{pmatrix} \phi_1 & 1 & 0 & \ldots & 0 \\ \phi_2 & 0 & 1 & \ldots & 0 \\ \ldots & \ldots & \ldots & \ldots & \ldots \\ \phi_{d-1} & 0 & 0 & \ldots & 1 \\ \phi_d & 0 & 0 & \ldots & 0 \end{pmatrix} \alpha_{t-1} + \begin{pmatrix} \theta_0 \\ \theta_1 \\ \ldots \\ \ldots \\ \theta_d \end{pmatrix} \eta_t \tag{4.13}$$

Here $\phi_j = 0$ if $j > d$ and $\theta_j = 0$ if $j > d$. **Notice there is no measurement noise.**

As an example consider a simple example of an ARMA(2,3) model

$$(1 - \phi_1 B + \phi_2 B^2)X_t = (1 + \theta_1 B + \theta_2 B^2 + \theta_3 B^3)\epsilon_t \tag{4.14}$$

In state space form this is

$$x_t = [1, 0, 0, \ldots, 0, 0]\alpha_t \tag{4.15}$$

with

$$\alpha_t = \begin{pmatrix} \phi_1 & 1 & 0 \\ \phi_2 & 0 & 1 \\ 0 & 0 & 0 \end{pmatrix} \alpha_{t-1} + \begin{pmatrix} \theta_0 \\ \theta_1 \\ \theta_2 \end{pmatrix} \eta_t \tag{4.16}$$

If we use a suffix to denote the components of the vector, so $\alpha_t' = [\alpha_t^{(1)}, \alpha_t^{(2)}, \alpha_t^{(3)}]$ we have

$$\alpha_t^{(1)} = \phi_1 \alpha_{t-1}^{(1)} + \alpha_{t-1}^{(2)} + \theta_0 \eta_t^{(1)} \tag{4.17}$$

$$\alpha_t^{(2)} = \phi_2 \alpha_{t-1}^{(1)} + \alpha_{t-1}^{(3)} + \theta_1 \eta_t^{(2)} \tag{4.18}$$

$$\alpha_t^{(3)} = \theta_2 \eta_t^{(3)} \tag{4.19}$$

From the last two equations we have

$$\alpha_t^{(2)} = \phi_1 \alpha_{t-1}^{(1)} + \theta_1 \eta_t^{(2)} + \theta_2 \eta_{t-1}^{(3)}$$

Hence if we substitute in the first equation 4.17 we have

$$\alpha_t^{(1)} = \phi_1 \alpha_{t-1}^{(1)} + \phi_2 \alpha_{t-2}^{(1)} + \theta_0 \eta_t^{(1)} + \theta_1 \eta_{t-1}^{(2)} + \theta_2 \eta_{t-2}^{(3)} \tag{4.20}$$

As we have been careful to ensure that *there is no measurement noise* $x_t = \alpha_t^{(1)}$ and so this gives the ARMA(2,3) model as stated.

4.1   General state space methods

As we saw above we have a measurement or observation equation

$$X_t = H\alpha_t + \epsilon_t \tag{4.21}$$

where H is a row vector of dimension d and ϵ is the measurement noise
   We also assume that the state variables satisfy

$$\alpha_t = \phi\alpha_{t-1} + K\eta_t \tag{4.22}$$

where ϕ is the $d \times d$ transition matrix and K is a $d \times n$ matrix of parameters. The state noise vector is an n-dimensional white noise vector whose covariance matrix is (obviously) diagonal and the terms on the diagonal are $\sigma_1^2, \sigma_2^2, \sigma_3^2, \ldots, \sigma_n^2$. The state noises and the measurement noise are uncorrelated. Using the (now matrix) recursion for the states $(I - \phi B)\alpha_t = K\eta_t$ we can substitute in the state equation to get

$$X_t = H[\phi^t\alpha_0 + \phi^{t-1}K\eta_1 + \phi^{t-2}K\eta_2 + \cdots + K\eta_t] + \epsilon_t \tag{4.23}$$

and when there is an infinite past

$$X_t = H(I-\phi B)^{-1}K\eta_t = HK\eta_t + H\phi K\eta_{t-1} + HK\phi^2\eta_{t-2} + \cdots + \epsilon_t \tag{4.24}$$

so we have an innovations form

$$X_t = (\Psi_0 + \Psi_1 B + \cdots)\eta_t + \epsilon_t \tag{4.25}$$

   If we examine the innovations form of the state space model given above it is clear that we need to control the powers of the transition matrix. In fact some quite subtle mathematics is required but we can show that a necessary and sufficient condition for a stationary model is that the eigenvalues of the transition matrix ϕ must lie **inside** the until circle. Other expressions, such as writing the covariances in model terms are possible but we leave these for the reader.

4.1.1   State representation and the likelihood

It is easy to see that the state space model we have discussed is just one of a family of equivalent models each giving the same outputs for the same noise inputs. Suppose we define $H^* = HT$ where T is some non-singular matrix. We can then define $\phi^* = T^{-1}\phi T$, $\alpha_t^* = T^{-1}\alpha_t$ and $K^* = T^{-1}K$. It follows that

$$X_t = HTT^{-1}\alpha_t + \epsilon_t = H^*\alpha_t^* + \epsilon_t \tag{4.26}$$

and

$$\alpha_t^* = T^{-1}\alpha_t = T^{-1}\phi\alpha_{t-1} + T^{-1}K\eta_t = \phi^*\alpha_{t-1}^* + K^*\eta_t \tag{4.27}$$

We can avoid problems of uniqueness by choosing transition matrices which are canonical. In fact we will choose structures which ensure that this is so but it is worth noting that if we do not specify the form of the transition matrix fairly tightly then our model may be just one of a class.

4.2 Estimation

Up to now we have skimmed over the details of the estimation of time series models. As most modellers use the techniques presented in their software package this is not unreasonable, however we will spend some time discussing the likelihood estimation of time series models via a state space formulation. To write down the likelihood we need to be able to express the joint density in terms of a product. $f(x_1, x_2, \ldots, x_t)$ in a more tractable form. Since

$$f(x_t|x_1, \ldots, x_{t-1}) = \frac{f(x_1, x_2, \ldots, x_t)}{f(x_1, \ldots, x_{t-1})} \tag{4.28}$$

This means that we may write the likelihood as

$$L(\theta) = \prod_{t=1}^{N} f(x_t|x_1, x_2, \ldots, x_{t-1}) \tag{4.29}$$

and in consequence the log of the likelihood is

$$\log L(\theta) = \sum_{t=1}^{N} \log f(x_t|x_1, x_2, \ldots, x_{t-1}) \tag{4.30}$$

Here θ is the vector of (unknown) parameters. Notice that the likelihood is a product of conditional distributions. We will as usual assume that these distributions are normal with means $E[x_t|x_{t-1}, \ldots, x_1]$ say $\mu(t|t-1)$. The log-likelihood is now

$$L(\theta) = \sum \left[\frac{1}{2}\log 2\pi - \frac{1}{2}\log F_t - \frac{1}{2F_t}(x_t - \mu(t|t-1))^2 \right] \tag{4.31}$$

$$= \frac{N}{2}\log 2\pi - \frac{1}{2}\sum_{t=1}^{N} \log F_t - \frac{1}{2}\sum_{t=1}^{N} \frac{V_t^2}{F_t} \tag{4.32}$$

Here $V_t = X_t - \mu(t|t-1)$ and F_t is the prediction error variance.

For any estimation methods based on the state space formulation of a model we need estimates of the state variables. Suppose that our estimate of α_t at time t is a_t while the estimate made at time t-1 is $a_{t|t-1}$. These estimates will have variance matrices

$$C_t = E[(\alpha_t - a_t)(\alpha_t - a_t)'] \tag{4.33}$$

$$C_{t|t-1} = E[(\alpha_t - a_{t|t-1})(\alpha_t - a_{t|t-1})'] \tag{4.34}$$

The beauty of the state space representation is that we have the Kalman filter updating equations (see Kalman (1960) or Janacek and Swift (1993)).
The Kalman filter recursions: we omit the derivations and just give the equations which split into

The prediction equations

$$a_{t|t-1} = \phi a_{t-1} \tag{4.35}$$

$$C_{t|t-1} = \phi C_{t-1} \phi' = K\Sigma K' \tag{4.36}$$

The updating equations

$$F_t = HC_{t|t-1}H' + \sigma_\epsilon^2 \tag{4.37}$$

$$C_t = C_{t|t-1} - \frac{1}{F_t}C_{t|t-1}H'HC_{t|t-1} \tag{4.38}$$

$$a_t = a_{t|t-1} + \frac{1}{F_t}C_{t|t-1}H'(x_t - Ha_{t|t-1}) \tag{4.39}$$

For constructing the likelihood

$$V_t = x_t - Ha_{t|t-1} \tag{4.40}$$

Given some starting values we step through the recursions and at each stage we obtain the prediction errors and the prediction error variances for any parameters set. This means we can compute the likelihood for any parameter set and with a suitable maximization procedure we can get maximum likelihood estimates. The Kalman approach provides compact computer code with the possibility of fast execution by comparison with alternative approaches. We use it to find our ARMA estimates as well as for the more overtly state space models.

The start up problem

If we start our recursions at time t=1 then we need the state estimate a_0 together with a covariance estimate C_0 at time zero. There are various possibilities, from using the unconditional expected values to rather exotic ones (see Harvey (1989)).

   We use the fact that the effect of the starting values is soon lost, especially with long series and we will set a_0 to zero and the covariance C_0 to M (a large number) times the unit matrix. This is a simple and effective technique which is widely used. It is also common for the recursions to settle down to a steady state when we have stationary series. By this we mean that C_t, $C_{t|t-1}$ and F_t converge to fixed (time independent) values. The advantage is that when this happens we can skip equations 4.36, 4.37 and 4.38. While there is no analytic result to tell us when this has happened we can put a numerical check in the recursions.

The reader may have noticed that we could have used a vector value of x_t in most of the algebra above. Indeed we can easily extend all our state space models to vector series, a point we will return to, briefly, in a later chapter.

4.2.1 Forecasting

We know that the minimum mean square error of the state α_t at any time is a_t. Thus, the one step ahead forecast at time t of x_{t+1} is $H\phi\alpha_t$ and the k ahead predictor is just $H\phi^k\alpha_t$. The variance of the forecast errors is

$$E[x_{t+k} - x_{t+k|t}]^2 =$$

$$E[(H\phi^k\alpha_t + H\phi^{k-1}\eta_{t+1} + H\phi^{k-2}\eta_{t+2} + \cdots + H\eta_{t+k} + \epsilon_{t+k} - H\phi^k a_t)^2]$$

As the noise terms are independent this becomes

$$H\left\{\phi^k C_t (\phi^k)' + \sum_{j=1}^{k} \phi^{k-j} K\Sigma K' \phi^{k-j})\right\} H' + \sigma_\epsilon^2$$

This then is the expression for the minimum mean square error of prediction.

Do bear in mind that this expression, and the explicit forms for the predictors assume that the correct model has been used, *this is most unlikely*, so any bound based on these expressions or indeed the earlier ones obtained in the ARIMA case will almost certainly be too narrow. They will however provide a useful guideline provided we are not too reliant on their actual magnitude.

4.2.2 Computation

We have used the state space form of the ARMA model together with the Kalman recursions to compute the likelihood and then to estimate model parameters. R r routines based on the state space model and the Kalman recursions are provided for the reader. They are slow since they are written in R but have the advantage of being reasonably transparent. There is also a variant of the E-M procedure of Shumway and Stoffer (1982). This is a possible alternative approach based on the E-M algorithm. Experiment shows that it runs in R and the reader may like to experiment with the routine.

We can of course work with state space models directly and this is our next step.

4.2.3 Simulation

The reader will no doubt have noticed that the state space representation is ideally suited to the generation of artificial series. If we specify the transition matrix ϕ and K then all we need is to set a value for the initial state and use randomly generated noise terms. We give a simple example in our R code. The results may be seen in Figure 3.3. Here we present the result of simulating a random walk model with drift. The equations are

$$x_t = \mu_t + \epsilon_t \tag{4.41}$$
$$\mu_t = \mu_t + \beta_{t-1} + \eta_t \tag{4.42}$$
$$\beta_t = \beta_{t-1} + \zeta_t \tag{4.43}$$

In matrix terms we have a vector of state variables $\alpha_t' = (\mu_t, \beta_t)$ with the recursion

$$\alpha_t = \begin{pmatrix} \mu_t \\ \beta_t \end{pmatrix} = \begin{pmatrix} 1 & 1 \\ 0 & 1 \end{pmatrix} \begin{pmatrix} \mu_{t-1} \\ \beta_{t-1} \end{pmatrix} + \begin{pmatrix} \eta_t \\ \zeta_t \end{pmatrix} \tag{4.44}$$

The R code is the function

```
simstate<-function(n,v)
{
 bign<-n+50
a<-c(0,0)
dim(a)<-c(2,1)
h<-c(1,0)
dim(h)<-c(1,2)
x<-seq(1,bign)

for(i in 1:bign)
{
x[i]<-h%*%a+rnorm(1)*v[1]
a<-phi%*%a
a[1,1]<-a[1,1]+rnorm(1)*v[2]
a[2,1]<-a[2,1]+rnorm(1)*v[3]
}
x[51:bign]
}
```

The parameters are the series length and the three noise variances held in the vector v. We generate $n + 50$ observations to allow for start up effects to dissipate. The six lines after the curly bracket set up the matrices, the loop then generates the series. Clearly the setting up of the matrices determines the type of the series, while the simple recursions just generate the series.

The same scheme, with suitably modified matrices can be used for the generation of ARMA models. The code looks very like the routines we have already developed using difference equations, which is of course hardly surprising.

4.2.4 Practical points

We have seen how the likelihood can be derived and to complete our discussion we look at examples of model fitting. While we will be looking at model fitting in

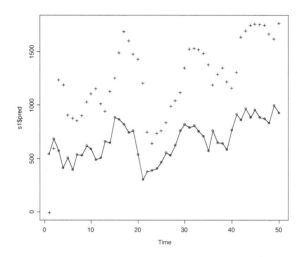

Fig. 4.1 *Artificial series and fitted (+) values*

rather more detail in the next chapter it will be helpful to run through an example.

We use an artificial series given in the following table. We try and fit a trending mean model to these data.

545.58	687.67	572.58	416.00	507.82	398.71	538.79	533.07	621.23
589.29	492.40	505.65	658.89	649.01	884.02	869.24	823.09	743.47
762.05	537.83	304.86	379.11	388.74	406.50	465.28	556.80	534.26
628.32	759.24	820.86	788.17	809.11	755.76	705.67	570.45	764.09
650.56	641.14	586.28	766.94	912.19	862.70	964.62	886.94	954.43
885.65	871.55	833.59	999.05	927.81				

A plot of the data and the predictions from the fitted model is given in Figure 4.1.

We set the initial state a_0 to zero and the initial covariance to 1000 times a unit matrix. Putting this into a wrapper and using the non-linear minimizer in R gives

```
nlm(space,c(1,1,1),print.level=2)
iteration = 0
Step:
[1] 0 0 0
Parameter:
[1] 1 1 1
Function Value
[1] 598446.4
```

```
Gradient:
[1] -3726971   1333559 -1196714

iteration = 1
Step:
[1]   780.5746 -279.2998   250.6391
Parameter:
[1]   781.5746 -278.2998   251.6391
Function Value
[1] 988.4778
Gradient:
[1] 0.1279468 0.0000000 0.1986971

iteration = 2
Parameter:
[1]   781.5746 -278.2998   251.6391
Function Value
[1] 988.4778
Gradient:
[1] 0.1279468 0.0000000 0.1986972

Successive iterates within tolerance.
Current iterate is probably solution.
```

In the above the first two parameters squared give the state noise variances while the third squared gives the measurement noise variances. As we can see from the plots of the predicted values and the residual the fit is not stunning, especially in the early stages, as we might expect. We could do better here by using some initial series values to give better starting values for the states.

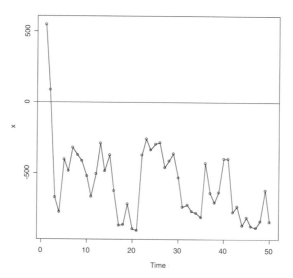

Fig. 4.2 *Residual after fitting trending mean to the artificial series*

The data series below is the global mean temperature in degrees Celcius for the years 1866 to 1985.

year	temp	year	temp	year	temp	year	temp	year	temp	year	temp	year	temp
1866	14.46	1886	14.55	1906	14.83	1926	15.09	1946	14.98	1966	14.95	1986	15.16
1867	14.36	1887	14.48	1907	14.53	1927	14.89	1947	15.15	1967	14.99	1987	15.27
1868	14.5	1888	14.69	1908	14.76	1928	15.06	1948	14.92	1968	14.93	1988	15.28
1869	14.6	1889	14.86	1909	14.73	1929	14.79	1949	14.83	1969	15.05	1989	15.22
1870	14.52	1890	14.6	1910	14.76	1930	14.89	1950	14.86	1970	15.02	1990	15.39
1871	14.62	1891	14.57	1911	14.72	1931	15.07	1951	14.99	1971	14.92	1991	15.36
1872	14.84	1892	14.64	1912	14.65	1932	15.04	1952	15.03	1972	15	1992	15.11
1873	14.87	1893	14.62	1913	14.76	1933	14.88	1953	15.1	1973	15.11	1993	15.14
1874	14.49	1894	14.7	1914	15	1934	15.05	1954	14.91	1974	14.92	1994	15.23
1875	14.33	1895	14.68	1915	14.95	1935	14.9	1955	14.92	1975	14.92	1995	15.4
1876	14.54	1896	14.87	1916	14.72	1936	15.03	1956	14.86	1976	14.82	1996	15.32
1877	14.95	1897	14.87	1917	14.65	1937	15.14	1957	15.08	1977	15.11		
1878	15	1898	14.64	1918	14.68	1938	15.13	1958	15.07	1978	15.05		
1879	14.52	1899	14.82	1919	14.86	1939	14.92	1959	15.04	1979	15.09		
1880	14.69	1900	15.02	1920	14.77	1940	15.16	1960	14.98	1980	15.18		
1881	14.71	1901	15.04	1921	14.88	1941	15.13	1961	15.1	1981	15.29		
1882	14.52	1902	14.75	1922	14.92	1942	15.07	1962	15.1	1982	15.08		
1883	14.57	1903	14.67	1923	14.86	1943	15.1	1963	15.1	1983	15.24		
1884	14.36	1904	14.5	1924	14.87	1944	15.16	1964	14.78	1984	15.11		
1885	14.55	1905	14.68	1925	14.86	1945	14.85	1965	14.88	1985	15.09		

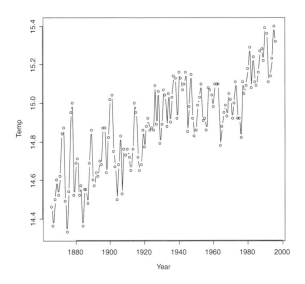

Fig. 4.3 *Annual global mean temperature*

We can fit our model to this series but it rapidly becomes apparent, as in Figure 4.3, that it is not appropriate (the fit is awful). The fitting of the 'correct' model and the choice of such models is the meat of our next chapter.

4.3 Exercises

1. Write the model

$$Y_t = \mu_t + \epsilon_t$$

$$\mu_t = \phi\mu_{t-1} + \eta_t$$

in ARIMA form.
2. Suppose $M_t = \alpha X_t + (1 - \alpha)M_{t-1}$, is used to produce forecasts. Is it possible to rewrite this exponential smoothing model in state space form?
3. Suppose

$$X_t = \mu_t + \epsilon_t$$

$$\mu_t = \mu_{t-1} + \beta_{t-1} + \eta_t$$

$$\beta_t = \phi\beta_{t-1} + \zeta_t$$

Find the autocorrelation function $\gamma(k)$ of the stationary part of the model. Show that this is equivalent to an ARIMA (p,d,q) model and give values for p, d and q.

4. Put the model in 3 in state space form and find the distribution of the state vector conditional on the first observation. Find an expression for $X_{t|t-1}$. What is the eventual form of the forecast function?

5
Fitting ARIMA models

We have spent some time discussing the behaviour of ARMA and ARIMA models without putting any of the methods to practical use. We now turn our attention to working with real data. There are two questions we must answer when presented with a series.

1. What is an appropriate model? The identification problem.
2. How do we estimate the parameters. The estimation problem.

We will treat these in reverse order.

5.1   The estimation problem

In one sense the problem of estimation is simple, given a model we compute the likelihood and then obtain the maximum likelihood parameters. The problems arise in the details. For the practical time series analyst the question becomes a pragmatic one. It is essential to have a programme to perform the estimation. Usually one has to use what is available and our experience is that most software is not based on the likelihood but on approximations to the likelihood.

These do work reasonably well and can be used with confidence especially if one uses common sense and a sceptical eye when looking at the diagnostics. Since we do not anticipate the reader devising their own routines we will restrict ourselves to an overview of the main approaches.

Likelihood methods

In many ways this is the simplest approach of all. We can show that the ARMA model can be written in state space form. The Kalman recursions then give us a way of calculating the likelihood. Maximizing the likelihood then gives a full likelihood-based approach. An alternative using the EM algorithm was detailed by Shumway (see the website detailed in the introduction for R code).

5.1.1 Conditional least squares

The state space approach to ARMA estimate is a relatively new one and most previous approaches look at variants of least squares. For a general ARMA model

$$X_t = \phi_1 X_{t-1} + \phi_2 X_{t-2} + \cdots + \phi_p X_{t-p} + \epsilon_t + \theta_1 \epsilon_{t-1} + \cdots + \theta_q \epsilon_{t-q}$$

we can write down the joint density of the noise terms, assuming that they are normal, as

$$\frac{1}{(2\pi\sigma^2)^{n/2}} \exp\left[-\frac{1}{2\sigma^2} \sum_{t=1}^{N} \epsilon_t^2 \right]$$

Then all we have to do is rewrite the model equation as

$$\epsilon_t = X_t - \phi_1 X_{t-1} - \phi_2 X_{t-2} - \cdots - \phi_p X_{t-p} + \theta_1 \epsilon_{t-1} + \cdots + \theta_q \epsilon_{t-q}$$

and substitution gives a log likelihood

$$-\frac{N}{2} log[2\pi\sigma^2] - \frac{1}{2\sigma^2} S(\Phi, \Theta, \sigma)$$

where here

$$S(\Phi, \Theta, \sigma) = \frac{1}{2\sigma^2} \sum_{t=1}^{N} \epsilon_t^2$$

While we can disentangle the noises we do have a problem at the start of the series. As we do not have an infinite past we need to make some assumptions about initial values. The simplest is to assume that all the noise terms up to time p are zero. In this case we have an approximation to the log-likelihood

$$-\frac{N}{2} log[2\pi\sigma^2] - \frac{1}{2\sigma^2} S^*(\Phi, \Theta, \sigma)$$

where $S^*(\Phi, \Theta, \sigma) = \frac{1}{2\sigma^2} \sum_{t=p+1}^{N} \epsilon_t^2$. This appears to be the most common form for computation. The problem we face is that we have lost data and our estimates are conditional on setting the initial values to zero. Box and Jenkins (1970) came up with a clever idea to overcome this problem. They suggested that one 'back forecast' the unknown initial values. We can illustrate this with an example. If we have the ARMA(1,1) model

$$(1 - \phi B)X_t = (1 + \theta B)\epsilon_t$$

This like any ARMA model may also be written in backwards form

$$(1 - \phi F)Y_t = (1 + \theta F)e_t \text{ where } F = B^{-1}$$

Since the model is stationary we deduce that the forward difference model will have exactly the same covariance structure as the difference model backwards.

We can also use the backward model to make 'forecasts' which give us estimates of the past values of the series. This technique, known as backcasting, gives us an improvement in estimation and removes the conditioning on the first few values of the series. The importance of backcasting increases for complex models, in particular for seasonal models.

Example 5.1. If we take the Wolfer sunspot series, Box series E and fit an AR2 model we have for approximate likelihood with backcasting from MINITAB

```
Estimates at each iteration
Iteration       SSE      Parameters
    0         108968    0.100    0.100    37.624
    1          90352    0.250    0.009    34.885
    2          74069    0.400   -0.083    32.206
    3          60064    0.550   -0.177    29.567
    4          48312    0.700   -0.271    26.959
    5          38795    0.850   -0.365    24.378
    6          31498    1.000   -0.460    21.821
    7          26409    1.150   -0.555    19.288
    8          23518    1.300   -0.651    16.776
    9          22792    1.409   -0.720    14.965
   10          22782    1.421   -0.727    14.755
   11          22782    1.422   -0.728    14.723
Relative change in each estimate less than  0.0010


Final Estimates of Parameters
Type        Estimate     St. Dev.   t-ratio
AR    1       1.4223      0.0721     19.74
AR    2      -0.7280      0.0718    -10.14
Constant     14.723       1.519      9.69
Mean         48.174       4.970


No. of obs.:   100
Residuals:    SS = 22363.8  (backforecasts excluded)
              MS =   230.6  DF = 97

Modified Box-Pierce chisquare statistic
Lag                  12           24            36            48
Chisquare  20.8(DF=10)  24.2(DF=22)  39.6(DF=34)  50.0(DF=46)
```

A full likelihood method in Splus gives

```
$model$ar:
[1]   1.5021585 -0.6116703
```

```
$model$ndiff:
[1] 0

$model$order:
[1] 2 0 0


$var.coef:
              ar(1)          ar(2)
ar(1)   0.006386321 -0.005952375
ar(2) -0.005952375  0.006386321

$method:
[1] "Maximum Likelihood"

$series:
[1] "wolfer"

$aic:
[1] 842.8445

$loglik:
[1] 838.8445

$sigma2:
[1] 305.4423
```

The R function `arpar3` gives

```
$params
[1]   1.5021095 -0.6116327

$varcov
              [,1]          [,2]
[1,]   0.006480141 -0.005951511
[2,] -0.005951511  0.006317310

$pred

 [1]  0.0000000 123.2178712  61.3980743  48.9853440  12.2060729
 [6] 25.1582493  -8.4458479  25.7607610 125.9614199 175.0546531
[11] 93.5722487  51.2252175  50.1546648  15.4891360  11.3064751
[16]  0.9535424  29.9343009 109.9959035 147.5129385 116.0408256
[21] 97.1250348  63.0171940  45.5943917  49.1471778  33.9011833
[26] 32.8397517   6.4673579  11.1894649  -0.7734665   2.3386417
```

```
[31]    8.0682356 16.7481040  42.5088650  46.7994151  37.0672361
[36]   45.8010491 33.7302285  16.3704917  -2.1046212   5.9005488
[41]   -1.8888428 -1.2232654   1.5021095   6.8989148  14.9671504
[46]   13.6899404 44.0109745  47.6898919  33.4513844  19.9863435
[51]   17.7016464  9.3545667   0.7286430   1.7270090   0.5576881
[56]   10.7936106 20.6427998  43.6781859  53.0866972  62.5491531
[61]   62.7201079 65.6703823  28.6753329  12.7006954  -5.1088402
[66]   14.6343618 77.6690163 148.3942943 132.6719194  70.3119631
[71]   66.1832469 42.0324846  17.0451901  13.4202173   1.8440192
[76]   15.8036826 50.9098893  68.6654803 109.2855026 126.3215717
[81]   68.3600547 40.4224859  55.7672485  41.9694189  25.5541036
[86]    7.6906234 -2.3295207   1.7270090  32.1019877  68.5484701
[91]  107.5584936 86.7090364  56.9456904  41.5287411  30.0064875
[96]   43.6873069 16.3165471   5.6847704   0.7286430  51.2966226
```

$perr
[1] 299.3513

$state
           [,1]
[1,]   74.00000
[2,] -22.63041

$like
           [,1]
[1,] 15068.18

$resid
  [1] -101.0000000   41.2178712   -4.6019257   13.9853440
  [5] -18.7939271   18.1582493 -28.4458479 -66.2392390
  [9] -28.0385801   50.0546531   8.5722487 -16.7747825
 [13]   12.1546648   -7.5108640   1.3064751 -23.0464576
 [17] -53.0656991 -22.0040965   16.5129385   -1.9591744
 [21]   7.1250348   -3.9828060 -14.4056083   2.1471778
 [25]   -7.0988167   11.8397517   -9.5326421   5.1894649
 [29]   -4.7734665   -4.6613583   -5.9317644 -17.2518960
 [33]   -2.4911350   3.7994151 -10.9327639   3.8010491
 [37]   5.7302285   6.3704917 -10.1046212   3.9005488
 [41]   -1.8888428   -2.2232654   -3.4978905   -5.1010852
 [45]   0.9671504 -21.3100596   -1.9890255   6.6898919
 [49]   3.4513844   -4.0136565   1.7016464   2.3545667
 [53]   -3.2713570   -0.2729910   -7.4423119   -6.2063894
 [57] -15.3572002   -6.3218141   -8.9133028   -4.4508469
 [61]   -8.2798921   17.6703823   0.6753329   4.7006954
 [65] -18.1088402 -42.3656382 -44.3309837   10.3942943
 [69]   29.6719194 -15.6880369   3.1832469   5.0324846
 [73]   -6.9548099   2.4202173 -13.1559808 -24.1963174
 [77] -11.0901107 -29.3345197 -14.7144974   30.3215717
 [81]   2.3600547 -23.5775141   1.7672485   2.9694189
 [85]   4.5541036   0.6906234   -6.3295207 -21.2729910
 [89] -22.8980123 -25.4515299   11.5584936   9.7090364
 [93]   -2.0543096   -2.4712589 -16.9935125   13.6873069
 [97]   0.3165471   -1.3152296 -36.2713570 -22.7033774

$ssq
[1] 41833.26
```
```

This function uses the matrix of second order derivatives (the Hessian) to compute the variance covariance matrix of the parameter estimates. It is clearly very similar to the Splus function. For more details of the output see Chapter 10.

5.2 Identification

Parameter identification is all very well but we first have to identify the model. There are some tools to help us in this.

- It is hard to see how we can expect to do very much without plotting the series. A plot will give us some clue as to whether there is a trend and that the series is non-stationary. Of course this is only an indication, what we see as a trend may be part of a very long-term cycle!
- The autocorrelations. If we have an MA(q) process the autocorrelations are zero after lag q while for an AR(p) process they decay exponentially. For a mixed ARMA(p,q) model we expect the correlations to tail off after lag (p-q).

  We must also bear in mind that we are using estimates of the autocovariances and autocorrelations, either of the form

  $$\hat{\gamma}(k) = \frac{1}{N} \sum_{s=1}^{N-|k|} (X_s - \bar{X})(X_{s+|k|} - \bar{X})$$

  or the unbiased form

  $$\hat{\gamma}(k) = \frac{1}{N - |k|} \sum_{s=1}^{N-|k|} (X_s - \bar{X})(X_{s+|k|} - \bar{X})$$

  We prefer the simplicity of the first. The distribution theory for these estimates is complex and not very useful. The *most useful result is that if $\gamma(k)$ is zero then $\hat{\gamma}(k)$ is approximately normal with variance $\frac{1}{N}$.*

5.2.1 The partial autocorrelations

We really need a bit more than this and the partial autocorrelations are a valuable tool. Suppose we decide that the model is AR(k) and fit

$$X_t = \phi_{1,k} X_{t-1} + \phi_{2,k} X_{t-2} + \cdots + \phi_{k,k} X_{t-k} + \epsilon_t$$

If we fitted k+1 terms we expect the coefficient will be slightly different so we have two subscripts, one of which denotes the length of the model. If we examine the last coefficient $\phi_{k,k}$ for $k = 1, 2, 3, \ldots$ we would expect that for an AR(p) model they will be approximately zero after the pth term. An MA(q) model however is usually an infinite AR model so the partial autocorrelations will just decay away. We can summarize the behaviour in Table 5.1.

Table 5.1 *Behaviour of the auto and partial correlations*

Process	Autocorrelation	Partial
AR(p)	Exponential decay or damped cosine	Zero after lag p
MA(q)	Cuts off after lag q	Exponential decay or damped cosine
ARMA(p,q))	Exponential decay after lag (q-p)	Decay after lag (p-q)

5.2.2 The corner point

We cannot disguise the fact that when one works with time series the above suggestions prove a very poor guide. There is perforce a good deal of trial and error. Remember we are seeking a model with small values of p and q; even so it is possible to do a good many trials with very many errors. There have been many suggestions for other additional identification tools, one of which is quite effective and is the Corner method of Gourieroux *et al.* (1980). They define the determinant

$$\Delta(i, j) = \det \begin{pmatrix} \rho(j+1) & \rho(j+2) & \cdots & \rho(j+i+1) \\ \rho(j) & \rho(j+1) & \cdots & \rho(j+i) \\ \cdots & \cdots & \cdots & \cdots \\ \rho(j+1-i) & \rho(j+2-i) & \cdots & \rho(j+1) \end{pmatrix} \quad (5.1)$$

They show that for an ARMA(p,q) process

1. $\Delta(i, j) = 0$ for all $i \geq p$ and $j \geq q$
2. $\Delta(i, q - 1) \neq 0$ for all $i \geq p - 1$
3. $\Delta(p - 1, q) \neq 0$ for all $j \geq q - 1$

If we estimate the determinants we should be able to detect the orders p and q by looking at an i j table of determinants. Consider a (large) table whose ijth element is $\Delta(i, j)$. The top left hand corner will be non zero but at some column the table becomes zero as we move right and as we move from top to bottom the table also becomes zero. If we look for the column which borders the zero region and the row which is also the border then we can pick out the AR and MA order. At least we hope so.

We have a simple R routine that does just that. It is probable that some degree of sophistication is worthwhile but the simple case appears quite effective.

Applying it to the purse data gives

```
       i/j    [0]            [1]            [2]            [3]            [4]
[0]    0.49283753   5.342152e-01    3.629771e-01    2.944238e-01    0.26092602474
[1]   -0.29132633   1.064971e-01   -2.553329e-02   -8.024806e-03    0.02006506694
[2]    0.00860501   6.746949e-03    5.860380e-04    9.626138e-05    0.00012552094
[3]    0.01199322   1.084453e-04   -1.169852e-05   -7.978591e-06   -0.00019082864
[4]    0.01024457   5.530129e-06    5.708480e-08    5.675102e-07    0.00004661894
```

From the above there seems to be a break at j=0 or j=1 and at i=1 or 2. The implication seems an AR(2) model. It seems sensible to try the scaled table, where we scale the entries by their standard deviation. In this case we have

i/j	[0]	[1]	[2]	[3]	[4]
[0]	2.94	3.188	2e+00	2e+00	2e+00
[1]	1.74	0.636	2e-01	5e-02	1e-01
[2]	0.05	0.044	2e-02	1e-02	2e-03
[3]	0.07	0.001	7e-04	2e-03	3e-03
[4]	0.06	0.001	8e-05	6e-05	7e-04

Again we are led to an ARMA(1,1) model.

For the Wolfer sunspot data we try the corner plot but here we use the determinants scaled by their standard deviation

i/j	[0]	[1]	[2]	[3]	[4]
[0]	4.07757	2.16514	4e-01	-0.85686	-1e+00
[1]	1.12238	0.64306	2e-02	0.14517	2e-01
[2]	0.02980	-0.00804	-4e-05	0.00436	7e-03
[3]	0.00471	0.00010	5e-06	0.00025	-9e-05
[4]	0.00002	0.00007	3e-08	0.00002	1e-06

There is a cut in the AR direction of 1 or 2 while there seems to be some indication of an order 1 moving average effect.

5.3 Model choice

We have seen how to estimate parameters and we have some guidelines towards identification of models. How can we decide between models and how can we see if the fit of a particular model is adequate? There are some simple criteria, one of which is the obvious one.

- Is the structure of the model reasonable? For example are the forms of the long range forecasts sensible? Is it possible that the model can be reduced in form by removing common terms?
- If we fit the correct model we expect the residuals, $e_t = X_t - \hat{X}_{t-1}$, to behave like white noise. Of course they will not be exactly noise even if we do have the right model but they should be a reasonable approximation since with the right model we expect zero autocorrelations.

  We should plot the residuals, examine them for patterns and also for outliers! It is usually best to standardize them and we can use the prediction error estimate to do this job.

- Since the residuals form a time series we can examine their autocorrelations etc. For noise we can show that the

$$var(\hat{\rho}_e(k)) \simeq \frac{1}{N}$$

where N is the series length, and as we stated above, we expect small correlations.

- If we are unsure it is always possible to fit a model to the residuals! Suppose we fit

$$\phi(B)X_t = \theta(B)a_t$$

but discover that we can fit a model, say

$$\phi^*(B)a_t = \theta^*(B)\epsilon_t$$

to the residuals. Then the correct model is

$$\phi(B)\phi^*(B)X_t = \theta^*(B)\epsilon_t$$

- It can be illuminating to *overfit* by fitting a model with an extra AR or MA term. Of course we also have the standard deviations of our parameter estimates which we can use to discard terms which seem inappropriate.

5.3.1   The portmanteau test

Another way of using the residuals is the portmanteau test. This is a popular test which exists in several slightly different forms. The original, proposed by Box and Pierce (1970) takes the form

$$Q_b = N \sum_{j=1}^{K} \hat{\rho}(j)^2$$

where the correlations are those of the residual series. The commonly used form is

$$Q = N(N+2) \sum_{j=1}^{K} \frac{1}{n-j}\hat{\rho}(j)^2$$

For $K \geq 15$ it has been shown that Q has a chi-squared distribution with K−p−q degrees of freedom. We would reject the hypothesis that the residual series is white noise if Q exceeds the percentage point of the chi-squared distribution.

The global mean temperature

We take as a further example the global mean temperature series from Chapter 4. A plot of the series shows that it appears to be increasing (See Figure 5.1) and is in consequence non-stationary. This is confirmed by the autocorrelation plot (Figure 5.2).

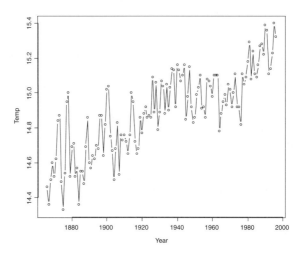

Fig. 5.1 *Global mean temperature by year*

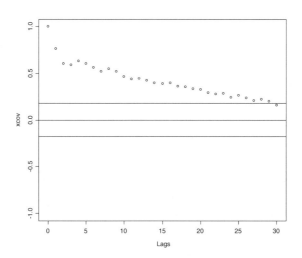

Fig. 5.2 *Autocorrelation plot for global mean temperature*

The autocorrelations are not dying away – typical behaviour for series which are not stationary. If we difference them the autocorrelation and partial autocorrelations imply an ARMA model.

The corner plot gives

```
         [0]    [1]    [2]   [3]    [4]
[0]   -1.7  -2.45 -0.841 1.168 2e-01
[1]    2.8   0.56  0.437 0.188 7e-03
[2]    2.1   0.37  0.102 0.033 2e-02
[3]    0.7   0.13  0.004 0.008 3e-03
[4]    0.5   0.05  0.009 0.002 8e-04
```

This appears to suggest, at least to the author that we might try an AR(2,1) model. Using arpar3 with the first 7 startup values omitted gives

```
$params
[1]   0.1967266 -0.3009533 -0.6912093

$varcov
                [,1]            [,2]           [,3]
[1,]   0.0098431875   0.0002064345 -0.003497853
[2,]   0.0002064345   0.0079087814 -0.001941104
[3,] -0.0034978534 -0.0019411042  0.004778290

$pred
  [1]   0.000000e+00 -1.396660e-02  5.751519e-02 -4.233991e-02
  [5]  -2.328301e-02 -3.634866e-02 -1.590645e-01 -1.892727e-01
  [9]   4.722474e-02  2.256957e-01  1.002987e-01 -1.964657e-01
 [13]  -2.838588e-01  2.607728e-02  7.842771e-02 -6.843175e-03
 [17]   8.320045e-02  8.996572e-02  1.509780e-01  7.360595e-02
 [21]  -6.304051e-03  3.025633e-02 -6.186107e-02 -1.900211e-01
 [25]  -5.394091e-02  5.579789e-02  1.298284e-02 -2.203238e-03
 [29]  -3.506240e-02 -3.842205e-02 -1.144902e-01 -1.363177e-01
 [33]   1.950689e-02 -6.304199e-03 -1.574255e-01 -1.788942e-01
 [37]   1.372754e-02  1.363236e-01  2.023663e-01  1.020326e-01
 [41]  -5.781808e-02  6.323728e-02  2.026525e-02 -4.037727e-02
 [45]  -3.371498e-02 -1.255340e-02  3.797486e-02 -7.077753e-03
 [49]  -1.566728e-01 -1.557983e-01  2.108944e-02  1.184102e-01
 [53]   8.807842e-02 -3.715481e-02 -3.534992e-02 -5.174141e-02
 [57]  -8.864825e-02 -4.364365e-02 -1.705449e-02 -9.852924e-03
 [61]  -1.175318e-01 -5.156181e-02 -5.951128e-02  4.121340e-02
 [65]   6.029623e-02 -7.742483e-02 -9.285383e-02  2.396436e-02
 [69]  -1.934506e-02  9.638785e-03 -1.247727e-02 -1.021414e-01
 [73]  -9.876103e-02  3.858630e-02 -2.880434e-02 -7.730413e-02
 [77]  -1.473576e-02 -6.962751e-03 -4.351024e-02  1.051576e-01
 [81]   1.016987e-01 -5.289087e-02  2.601020e-02  1.317011e-01
 [85]   1.032843e-01 -1.920279e-03 -6.023052e-02 -8.828374e-02
```

```
 [89]  1.186239e-02  6.043568e-02  6.843306e-02 -4.342734e-02
 [93] -9.128225e-02 -4.525109e-02  7.419580e-03 -3.615218e-02
 [97] -6.110309e-02 -4.223499e-02  1.290411e-01  1.360511e-01
[101]  2.933066e-02 -2.057241e-02  3.410970e-03 -3.892296e-02
[105] -4.818382e-02  2.517173e-02  7.935679e-03 -7.298408e-02
[109]  1.039951e-02  6.436935e-02  9.394086e-02 -4.837174e-02
[113] -9.104249e-02 -6.465144e-02 -1.012292e-01 -1.514493e-01
[117] -3.394668e-02 -3.938119e-02 -1.109047e-02  4.134774e-02
[121] -1.476803e-05 -7.546997e-02 -9.021517e-02 -3.569812e-02
[125] -9.067961e-02 -9.900612e-02  6.421523e-02  1.047900e-01
[129]  1.889975e-02 -9.808408e-02
```

$perr
[1] 0.01665411

$state
```
            [,1]
[1,] -0.08000000
[2,] -0.06366192
```

$like
```
        [,1]
[1,] 128.8795
```

$resid
```
 [1]  0.100000000 -0.153966598 -0.042484806  0.037660085
 [5] -0.123283015 -0.256348660 -0.189064534  0.190727265
 [9]  0.207224736  0.015695688 -0.309701286 -0.246465682
[13]  0.196141245 -0.143922723  0.058427706  0.183156825
[17]  0.033200448  0.299965721 -0.039021953  0.073605947
[21]  0.063695949 -0.179743668 -0.231861073  0.069978950
[25] -0.023940906 -0.014202112  0.032982835 -0.082203238
[29] -0.015062400 -0.228422053 -0.114490189  0.093682270
[33] -0.160493115 -0.206304199 -0.177425523  0.111105821
[37]  0.093727535  0.306323604  0.022366273 -0.047967401
[41]  0.242181916 -0.166762721  0.050265251 -0.070377270
[45]  0.006285016  0.057446597 -0.072025144 -0.247077753
[49] -0.106672765  0.074201750  0.091089444  0.088410202
[53] -0.091921577  0.052845195 -0.145349921 -0.091741409
[57] -0.028648253 -0.053643645 -0.007054491 -0.239852924
[61]  0.082468224 -0.221561812  0.210488724 -0.058786598
[65] -0.119703769 -0.047424826  0.067146166 -0.146035642
[69]  0.130654937 -0.120361215 -0.122477272 -0.092141352
[73]  0.111238975 -0.201413704  0.001195664 -0.017304126
[77] -0.044735758 -0.066962751  0.266489763 -0.024842399
[81] -0.068301317  0.177109131  0.116010201  0.101701114
[85] -0.026715718 -0.041920279 -0.130230521  0.101716265
[89]  0.001862387  0.120435683 -0.151566939 -0.033427341
```

```
[93]  -0.061282250   0.014748913  -0.112580420  -0.036152178
[97]  -0.061103090   0.277765015   0.029041083   0.066051148
[101] -0.010669340   0.039427594  -0.116589030  -0.008922964
[105]  0.051816182  -0.054828265  -0.102064321   0.117015919
[109]  0.010399511   0.164369351  -0.196059137   0.011628257
[113] -0.131042487  -0.154651445  -0.211229158   0.058550705
[117] -0.193946684   0.090618807   0.008909529  -0.028652264
[121] -0.110014768  -0.085469966  -0.030215174  -0.205698117
[125] -0.060679609   0.150993878   0.034215232   0.014789976
[129] -0.151100246  -0.018084076
```

$ssq
[1] 2.203150

terms	ϕ_1	ϕ_2	θ_1
coeffs	0.1967266	−0.3009533	−0.6912093
S.errors	0.0992	0.0889	0.0691

The ARMA(2,1) model fitted to the differenced series, that is an ARIMA(2,1,2) model to the whole series, seems a reasonable model. The histogram and QQ plot of the residuals in Figure 5.3 do not appear to indicate any departures from randomness.

And neither does the plot of the residuals in Figure 5.4.

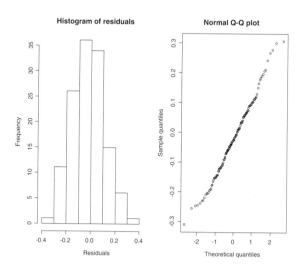

Fig. 5.3 *Residual plots for global mean temperature*

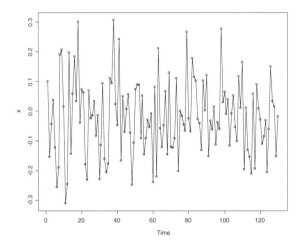

Fig. 5.4 *Residual plot for global mean temperature*

The portmanteau statistic gives

statistic	degrees of freedom
23.311 86	31
49.881 32	62
84.524 49	93

while the 'z' statistics (estimate over standard error) for the parameters are

terms	ϕ_1	ϕ_2	θ_1
coeffs	0.196 7266	−0.300 9533	−0.691 2093
z statistics	2.054	−3.453	−9.263

The portmanteau statistic inclines us to the belief that the model fits reasonably well while the 'z' statistics imply that all of the estimated coefficients are non-zero. Looking at the autocorrelations of the residuals there is the suggestion of some structure so try fitting a further model to the residuals. An MA(1) seems appropriate and does give plausible residuals. This implies that our original model was not quite right and we could have done better with an extended MA part on the model. However if we overfit our original model the extra MA term is not significant in terms of testing nor does it make a great difference to the residuals. Since the two models are very close we will pick the simpler and settle on our original model. The plot of the fitted versus the predicted is shown in Figure 5.5.

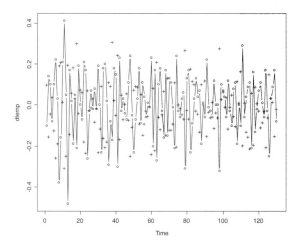

Fig. 5.5 *Original and predicted values for global mean temperature*

5.3.2   Model comparison criteria

It is not uncommon to have two models and to be unable to distinguish between them using the methods above. To overcome this difficulty Akaike suggested a criterion based on information. This Akaike information criterion AIC is widely used and while others have been suggested it is still the most popular. All we do is choose the model which minimizes the criterion. Two of the most popular criteria are

1. The Akaike Information Criterion
   $AIC(p,q) = ln\hat{\sigma}^2 + 2(p+q)/n$
2. The Bayes Information Criterion
   $BIC(p,q) = ln\hat{\sigma}^2 + 2(p+q)ln(n)/n$

In both cases $\hat{\sigma}^2$ denotes the variance of the residuals.

As an example, suppose one wished to fit the 'best' AR model to a series. The one possibility is to fit larger and larger order models until the AIC is a minimum. The minimum AIC model is then the best. The estimate of σ^2 is obtained from the Yule–Walker equations

$$\sigma^2 = var(X_t)[1 - \gamma(1) - \gamma(2) - \cdots - \gamma(p)]$$

A simple but effective procedure is to fit models in sequence and to choose the one with minimum AIC. For example if we take the Lynx series, or rather the log Lynx series we use an R function `stepar2` to fit a series of models and to compute the AIC. The results can be seen plotted in Figure 5.6.

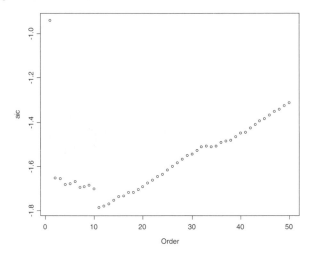

Fig. 5.6 *Plot of AIC against AR order*

As is apparent the minimum AIC is at order 11 and the partial autocorrelations are now

```
 [1]   0.78064036 -0.68272468 -0.21015008 -0.16832887  0.05994797
 [6]   0.09536027  0.18703284  0.10339262  0.10520238 -0.16347756
[11]  -0.26254459
```

Sadly just plotting the innovation variance is not sufficient as it is usually very difficult to assess the model order based on just $\hat{\sigma}^2$. While we favour AIC there are various other criteria. Akaike first suggested the final prediction error criterion FPE where

$$FPE(p) = \hat{\sigma}^2 \left(\frac{N + p + 1}{N - p - 1} \right)$$

while Rissanen proposed the minimum description length MDL

$$MDL(p) = N\hat{\sigma}^2 + p\log(N)$$

At present it is not clear which of the above, or the several we have not mentioned, is the optimum. Our preference is for AIC as it is widely used and appears to work well. As confirmation we try the sunspot series which we expect to be AR(2). AIC gives the results in Figure 5.7 which fit with our beliefs in earlier chapters.

Our experience is that on average a series without a clear AR model requires some N/3 terms to produce a reasonable AR model for the data. We use the recursive procedure for the partial autocorrelations to get the coefficients in a stepwise fashion.This is a reasonably quick and simple operation which does not need a matrix inversion. Interested readers can try the code on their own series.

We end by pointing out that fitting models is a skill which can be sharpened with practice. It is important to look carefully at each fit as this is the key to the whole procedure.

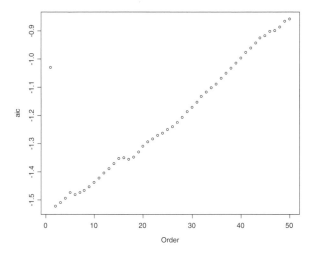

Fig. 5.7 *Plot of AIC for AR models of increasing order*

5.4 Seasonal series

It is clear that in the real world many series have seasonal fluctuations and we have said very little about modelling such series. How to begin? If we think in terms of difference equations then $(1 - B^s)X_t = 0$ will have solutions which will involve complex roots (the s roots of unity) and these will imply a trigonometric expression which has period s. Or forgetting the mathematics, looking at the changes over a seasonal cycle s would seem a good idea. Indeed there is some force to the idea of modelling points in the series which are s apart, an ARMA model for seasonal values. If we add this to the standard ARIMA model we have a seasonal model of the form

$$(1 - B)^d(1 - B^s)^D\phi(B)\Phi(B^s)X_t = \theta(B)\Theta(B^s)\epsilon_t \qquad (5.2)$$

where $\Phi(B^s)$ and $\Theta(B^s)$ are polynomials in B^s introduced to model the seasonal effects. This is often written as a ARIMA (p,d,q)×(P,D,Q) model where P and Q are the orders of the seasonal polynomials.

To fit such models and to identify them is of course a more difficult task than for the common non-seasonal ARIMA model. The available tools are the same and the modelling ideas carry forward in exactly the same way. It is just more difficult!

As an example we look at monthly Australian beer consumption which has aggregated into quarters from March 1956 until June 1994. The data are shown in Table 5.2 (read by rows). The series is not stationary (plot it to see) but the differences show promise as can be seen from Figure 5.8.

Table 5.2 *Beer consumption in Australia in mega litres*

284.4	226.9	308.4	262.0	227.9	236.1	320.4	271.9
232.8	237.0	313.4	261.4	226.8	249.9	314.3	286.1
226.5	260.4	311.4	294.7	232.6	257.2	339.2	279.1
249.8	269.8	345.7	293.8	254.7	277.5	363.4	313.4
272.8	300.1	369.5	330.8	287.8	305.9	386.1	335.2
288.0	308.3	402.3	352.8	316.1	324.9	404.8	393.0
318.9	327.0	442.3	383.1	331.6	361.4	445.9	386.6
357.2	373.6	466.2	409.6	369.8	378.6	487.0	419.2
376.7	392.8	506.1	458.4	387.4	426.9	565.0	464.8
444.5	449.5	556.1	499.6	451.9	434.9	553.8	510.0
432.9	453.2	547.6	485.8	452.6	456.6	565.7	514.8
464.3	430.9	588.3	503.1	442.6	448.0	554.5	504.5
427.3	473.1	526.2	547.5	440.2	468.7	574.5	492.6
432.6	479.8	575.7	474.6	405.3	434.6	535.1	452.6
429.5	417.2	551.8	464.0	416.6	422.9	553.6	458.6
427.6	429.2	534.2	481.7	416.0	440.2	538.7	473.8
439.9	446.8	597.5	467.2	439.4	447.4	568.5	485.9
442.1	430.5	600.0	464.5	423.6	437.0	574.0	443.0
410.0	420.0	532.0	432.0	420.0	411.0	512.0	449.0
382.0							

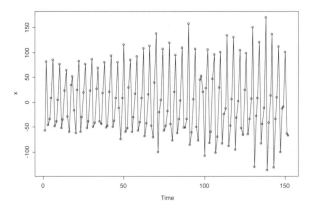

Fig. 5.8 *Differences of beer consumption by quarter*

The autocorrelations and partial autocorrelations in Figure 5.9 show a strong quarterly effect and indicate that some autoregressive component is required for any potential model.

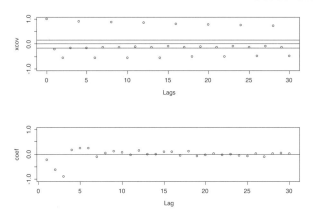

Fig. 5.9 *Autocorrelations and partial autocorrelations of beer consumption*

We tried seasonally differencing once and then fitted an ARIMA(2,1,1) model. This seemed to understate the seasonal features of the data so we tried a SARIMA (2,1,0)× (1,1,1) model. This is not too bad a fit as can be seen from the plot of the residuals and their autocorrelation plot in Figure 5.10.

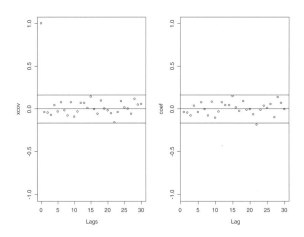

Fig. 5.10 *Autocorrelations and partial autocorrelations of beer consumption residuals*

The final model being

$$(1-B)(1-B)^4(1-0.922B-0.462B^2)(1+0.170B^4-0.139B^8)X_t = (1+0.723B^4)\epsilon_t$$

which has good portmanteau test results.

6

The frequency domain and the spectrum

6.1 Introduction

The spectrum gives us a way of studying time series which complements those used in the time domain. Suppose we have a series which is stationary to second order, say $\{X_t, t = \ldots, -2, -1, 0, 1, 2, \ldots\}$. That means that the mean $E[X_t] = \mu$ does not depend on time and the autocovariances $E[(X_t - \mu)(X_s - \mu)] = \gamma(t - s)$ depend only on $|t - s|$. While we can study the series using the autocovariances and the autocorrelations, as we have done in earlier chapters, a rather different approach is available via the spectrum.

In Chapter 1 we thought of series as having cycles. If we have a series of length N then we can imagine possible cycles whose periods can be 2,3,4, up to N. The smallest period we can detect is 2 because we cannot distinguish between 2 and shorter periods. We discuss this point in more detail under Aliasing. In terms of frequency (remember frequency is the reciprocal of the period) we can have cycles with frequencies $\frac{1}{2}, \frac{1}{3}, \ldots, \frac{1}{N}$. It is simpler to work in angular frequencies which are measured in radians per unit time rather than cycles per unit time. To get the angular frequencies we just multiply by 2π. For simplicity of notation for the moment we will write $\omega_j = \frac{2\pi}{N}$

Think of our stationary series $\{X_t\}$ as generated by a whole series of cycles

$$X_t = \sum_{j=1}^{p} [a_j \cos(\omega_j t) + b_j \sin(\omega_j t)] \tag{6.1}$$

where the a_j and b_j are zero mean sequences of independent random variables $j = 1, 2, \ldots, p$ having variances σ_j^2. This looks very much like a trigonometric regression model and is often known as a harmonic model. We can regard a model of this form as an attempt to explain the signal $\{X_t\}$ in terms of contributions at the 'frequencies' $\omega_1, \omega_2, \ldots, \omega_p$. A close analogy is that of a musical instrument; a note played on such an instrument is the sum of harmonic vibrations at different frequencies. Since we work with discrete series it is sufficient to work with frequencies in the range $-\pi$ to π since with a unit time interval the shortest period is 2 and the maximum (angular) frequency is π. Suppose that we assume for simplicity that $\omega_p = \pi$ then, since

$$\cos \omega + i \sin \omega = e^{i\omega}$$

we may rewrite 6.1 as

$$X_t = \sum_{j=-p}^{p} z_j e^{i\omega_j t} \tag{6.2}$$

where

$$z_{-j} = \frac{1}{2}(a_j + ib_j) \text{ and } z_j = \frac{1}{2}(a_j - ib_j) \text{ for } 0 < |j| < p$$

while at the end point $z_p = a_p$. The frequencies w_{-j} are to be taken as $-w_j$ and if we take the obvious step of writing z_{-j} as $-z_j$, the complex conjugate of z_j, we have the nice equivalent form in and the novel idea of a 'negative frequency'.

If we use this complex formulation then we can show that

$$\text{var}(X_t) = \sum_{j=-p}^{p} \sigma_j^2 \tag{6.3}$$

We see that the variance of the process is made up of contributions which are distributed over the individual frequencies. In the same way as with probability distributions we can think of a distribution of variance components (we shall use power as an abbreviation) and to this end we define the cumulative spectral function $H(\omega)$ as the sum of contributions to the series variance from components with frequencies less than or equal to ω. At the end points have $H(-\pi) = 0$ and $H(\pi) = \text{var}(X_t)$. Suppose now we imagine that the number of frequency points in our model becomes very large. Then the function $H()$ will become smoother until, in the limit, we will have a smooth function.

This is an exact analogy to the cumulative distribution function of a continuous random variable. Then we must think of the contribution to the total variance made in a frequency band rather than at a specific frequency, thus the band $[\omega, \omega + \delta\omega]$ contributes $H(\omega + \delta\omega) - H(\omega)$ to the total variation. Notice we have assumed that there is no dominant frequency contributing a large amount of power and the contribution from any point frequency is zero. If in this limiting case $H(\omega)$ is differentiable then the derivative $h(\omega)$ is defined as

$$H(\omega) = \int_{-\pi}^{\omega} h(\phi)d\phi \text{ or } h(\omega) = \frac{\partial H(\omega)}{\partial \omega}$$

and is called the power spectrum. This function $h(\omega)$ is the analogue of the familiar probability density function in statistics and gives the distribution of power over the frequency range. However it should be borne in mind that the H *need not have a derivative*. If a frequency 1 contributes a finite amount of power then the cumulative spectrum will have a jump at that frequency; indeed a major use of the spectrum is to locate the jumps and hence find the frequencies which give finite power. The connection with the covariance function can be established in the finite case as

$$\gamma(s) = \sum_{j=-p+1}^{p} e^{-i\omega_j t}[H(\omega_j) - H(\omega_{j-1})]$$

and following the same sort of limiting process as before we have, for a differentiable function H, the important relation between the power spectrum $h(\omega)$ and the autocovariance

$$\gamma(s) = \int_{-\pi}^{\pi} e^{-i\omega t} h(\omega) d\omega$$

and hence using the ideas of Fourier series we have the inverse relationship

$$h(\omega) = \frac{1}{2\pi} \sum_{s=-\infty}^{\infty} \gamma(s) e^{i\omega s}$$

and since $\gamma(s) = \gamma(-s)$ we have for real series

$$h(\omega) = \frac{1}{2\pi} \sum_{s=-\infty}^{\infty} \gamma(s) \cos(\omega s)$$

and the power spectrum is symmetric around $\omega = 0$. We summarize as follows: For any stationary time series we can define a cumulative power spectrum $H(\omega)$ defined on a set of frequencies $-\pi \leq \omega \leq \pi$ and

(a) The cumulative power spectrum $H(\omega)$ defines the amount of 'power' or the contribution to the total variance made by frequencies below ω.

(b) When $H(\omega)$ is continuous then there is a spectral density function $h(\omega)$ which satisfies

$$h(\omega) = \frac{1}{2\pi} \sum_{s=-\infty}^{\infty} \gamma(s) \cos(\omega s) \tag{6.4}$$

and

$$\gamma(s) = \int_{-\pi}^{\pi} e^{-i\omega t} h(\omega) d\omega \tag{6.5}$$

(c) Harmonic components with finite power produce jumps in $H(\omega)$ and consequently spikes or delta functions in $h(\omega)$. In most cases the function H will usually be continuous with possibly some jumps corresponding to sinusoids at a finite number of frequencies and the continuous case the power spectral $h(\omega)$ will satisfy 6.4.

While we have started from a simple harmonic model it is possible to be very much more general and show that any second order stationary series has a spectral distribution $H(\omega)$. This function can be thought of as being of the form $a_1 H_1(\omega) + a_2 H_2(\omega) + a_3 H_3(\omega)$ where $H_1(\omega)$ is a discrete function, $H_2(\omega)$ is continuous and $H_3(\omega)$ is a 'singular' component. Of course some of the coefficients may be zero. A typical plot of $H(\omega)$ is shown in Figure 6.1.

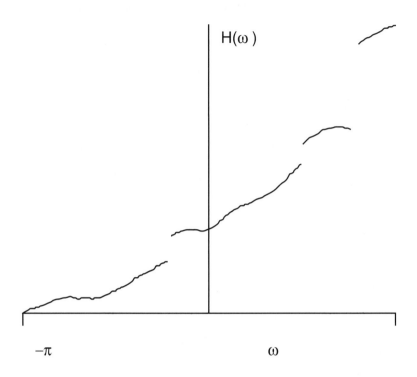

Fig. 6.1 *Examples of H(ω)*

6.1.1 Some examples of spectra

- White noise process. Since the autocovariances are zero except at lag zero then from 6.4 we have

$$h(\omega) = \frac{\sigma^2}{2\pi}$$

  This gives a flat spectrum.

- For a first order autoregressive model $X_t = \phi X_{t-1} + \epsilon_t$ where ϵ_t is white noise we have from the Yule–Walker equations $\gamma(k) = \phi\gamma(k-1)$ so using 6.4 we have, after some algebra

$$h(w) = \frac{1}{2\pi} \sum_{s=-\infty}^{\infty} \phi^s \gamma(0) = \frac{\gamma(0)}{2\pi}/(1+\phi^2 - 2\phi\cos\omega) \text{ for } -\pi < \omega < \pi$$

- For an MA(1) process, say $X_t = \epsilon_t + \theta\epsilon_{t-1}$ we know that the autocovariances are zero after lag one and that $\gamma(1) = \theta\sigma^2$ while $\gamma(0) = (1+\theta^2)\sigma^2$ so

$$h(\omega) = \frac{\sigma^2}{2\pi}(1 + \theta^2 + 2\theta\cos(\omega))$$

Four examples of spectra are shown in Figure 6.2.

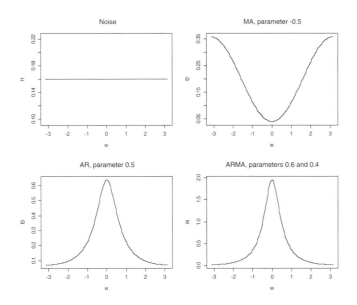

Fig. 6.2 *Examples of four spectra*

The flat white noise spectrum shows that the contributions to variance are equally distributed across the entire frequency range while for the AR model, here with parameter 0.5, the contributions are greater from the low frequency which is the long period end. This would imply that there are rather weak short period effects and in consequence we would expect a smoother series than white noise. The MA spectrum, again with a parameter of 0.5 shows the opposite effect, the power is concentrated towards the high frequency end of the frequency range and in consequence we expect strong high frequency, i.e. short period effects giving an irregular appearance to a realization generated by such a model.

A further example

At a further example take the AR(2) model $X_t = \frac{3}{2}X_{t-1} - \frac{15}{16}X_{t-2} + \epsilon_t$ which has been chosen to exhibit pseudo cyclic behaviour. The spectrum is given in Figure 6.3.

From the graph we see there is a large contribution to the power from frequencies around f = 0.11 cycles per unit time indicating behaviour which will look cyclic. However it is not a true cycle as the spectrum $H(\omega)$ does not have a jump and the peak in $h(\omega)$ while large is still a peak and not a singular point.

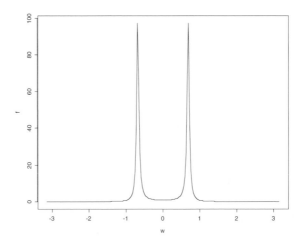

Fig. 6.3 *Spectrum of an AR(2) with pseudo-cyclic behaviour*

6.2   The spectral representation

Looking back we have written the series as a sum of complex random variables of the form

$$X_t = \sum_{-p+1}^{p} z_j e^{i\omega_j t}$$

One might ask what happens to this as the number of individual frequencies becomes infinite? We can handle this by defining a stochastic process $Z(\omega)$ which is in some sense the accumulation of the z_j's up to frequency ω. While the mathematics is rather complex the new process is rather interesting for our purposes. We will avoid the details and just think of $Z(\omega)$ as a complex process related to z_j via $z_j = Z(\omega_j) - Z(\omega_j - d\omega_j)$. It can be shown that for non-overlapping intervals $[\omega_2, \omega_1]$ and $[\omega_4, \omega_3]$

$$E[Z(\omega_2 - \omega_1)\bar{Z}(\omega_4 - \omega_3)] = 0 \qquad (6.6)$$

For most purposes we just use the rather more simple relationships

$$E[dZ(\omega)d\bar{Z}(\theta)] = \begin{cases} 0 & \text{for } \omega \neq \theta \\ h(\omega)d\omega & \text{when } \omega = \theta \end{cases}$$

We can thus take the limit as

$$X_t = \int_{-\pi}^{\pi} e^{i\omega t} dZ(\omega) \tag{6.7}$$

where the process $Z(\omega)$ satisfies equation 6.6.

The process $Z(\omega)$ is known as a Wiener Process. While the full mathematical structure is difficult to derive, the formula above, known as the spectral representation of all stationary processes gives us a simple way to deal with the calculation of the spectra of linear models. It also gives insight into the very important idea of filters as we shall see.

6.3 Linear filters

One very important reason for the utility of the spectral form of time series analysis is its value when studying linear transformations on time series. Suppose we have a series $\{X_t\}$ which passes into a 'black box' which produces $\{Y_t\}$ as output, rather as in the schematic in Figure 6.4.

Fig. 6.4 *Input and output of a black box*

We could regard the effect of the box as an operation on the input giving output, say $Y_t = LX_t$. This would model a very large number of common situations, for example the impact of a bump on the road is modified by the suspension system to give an output to the car occupants. While the model is so general it is difficult to study in detail so we make two crucial restrictions

1. That the relationship is linear
2. That the relationship is invariant over time.

It is not obvious these restrictions mean in effect that for any t, Y_t is a weighted linear combination of past and future values of the input viz.

$$Y_t = \sum_{j=-\infty}^{\infty} a_j X_{t-j} \tag{6.8}$$

with $|\sum a_j| < \infty$, for technical details see for example Koopmans (1974). We call the relationship in equation 6.8 a linear filter and much of time series analysis is based on the study of such filters. While we can study filters in the time domain it is often very much easier to look at the relationship between the input and output spectra. Suppose the input series $\{X_t\}$ has a power spectrum $h_x(\omega)$ and the output $\{Y_t\}$ a corresponding spectrum $h_y(\omega)$ then we can use the spectral representation as follows.

$$Y_t = \int_{-\pi}^{\pi} e^{i\omega t} dZ_y(\omega) = \sum_{j=-\infty}^{\infty} a_j X_{t-j}$$

$$= \sum_{j=-\infty}^{\infty} a_j \int_{-\pi}^{\pi} e^{i\omega(t-j)} dZ_x(\omega)$$

so

$$Y_t = \int_{-\pi}^{\pi} e^{i\omega t} dZ_y(\omega) = \int_{-\pi}^{\pi} e^{i\omega(t)} \sum_{j=-\infty}^{\infty} a_j e^{-i\omega j} dZ_x(\omega)$$

thus

$$dZ_y = \sum_{j=-\infty}^{\infty} a_j e^{-i\omega j} dZ_x(\omega)$$

and so

$$f_y(\omega) = \left| \sum_{j=-\infty}^{\infty} a_j e^{-i\omega j} \right|^2 f_x(\omega)$$

In consequence it is very easy to relate the input and output spectra of a linear filter, for the relation

$$Y_t = \sum_{j=-\infty}^{\infty} a_j B^{-j} X_t$$

gives

$$f_y(\omega) = |\Gamma(\omega)|^2 f_x(\omega) \tag{6.9}$$

where $\Gamma(\omega) = \sum_{j=-\infty}^{\infty} a_j e^{-i\omega j}$. The function $\Gamma(\omega)$ is called the transfer function or the frequency response function while its modulus $|\Gamma(\omega)|$ is often called the amplitude gain. The squared value $|\Gamma(\omega)|^2$ is known as the gain or the power transfer function of the filter. The argument $arg(\Gamma(\omega))$ is the phase gain or just the phase. There is rather a rich variety of nomenclature since filters are widely used in many fields, especially in engineering. Suppose we apply a moving average to a series, say

$$Y_t = \frac{1}{5}(X_{t+2} + X_{t+1} + X_t + X_{t-1} + X_{t-2}) = \frac{1}{5}(B^2 + B + 1 + B^{-1} + B^{-2})X_t$$

then $5\Gamma(\omega) = e^{-2i\omega} + e^{-i\omega} + 1 + e^{i\omega} + e^{2i\omega} = 1 + 2\cos\omega + 2\cos 2\omega = \sin(5\omega/2)/\sin(\omega/2)$ and as we can easily see this function is zero when $5\omega/2 = 2k\pi$ or $\omega = 4k\pi/5$ $\quad k = \ldots -2, -1, 1, \ldots$ If this filter is applied to a series with a cycle of period 5, i.e. frequency $2\pi/5$, the resulting output spectrum given by $h_y(\omega) = |\Gamma(\omega)|^2 h_x(\omega)$ will have a zero at this frequency and in consequence the output series will not contain this cyclic effect. The gain is plotted in Figure 6.5. It should also be noted that in addition to having zeros which remove cycles the gain gets smaller as the frequencies increase. This means that the output series, compared to the input, has less power at high frequencies and in consequence the output series will have smaller high frequency effects and will appear smoother.

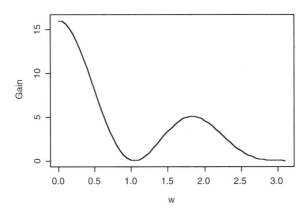

Fig. 6.5 *Filter gain*

6.3.1 Filter design

The design of filters is of considerable importance and much attention has been paid to the problems involved. Many filters are interesting because of the way they act in the frequency domain. High pass filters remove low frequencies while low pass filters do the opposite. To construct a filter with a prescribed transfer function can be very difficult. Suppose we wanted a filter whose transfer function was

$$\Gamma(\omega) = |\omega|$$

Then we need a trigonometric function equal to $\Gamma()$ in fact

$$\Gamma(\omega) = \sum_{j=0}^{m} c_j e^{-ij\omega}$$

for some set of coefficients $\{c_j\}$. One possibility is to find the Fourier series for $\Gamma(\omega)$ and use a truncated version as the filter function.

Thus for our example

$$\Gamma(\omega) = \frac{\pi}{2} - \frac{4}{\pi} \sum_{j=0}^{\infty} \frac{\cos[(2j+1)\omega]}{(2j+2)^2}$$

and we might take the first four terms. This works reasonably well except when there are sharp edges to the transfer function. As an illustration, in Figure 6.6 we show the filter together with three approximations using 1, 2 and 5 cosine terms. If we have two terms, then

$$\Gamma(\omega) = \frac{\pi}{2} - \frac{4}{\pi} \frac{\cos[\omega]}{4}$$

which we can decipher as

$$\Gamma(\omega) = \frac{\pi}{2} - \frac{1}{\pi} \frac{e^{\omega} + e^{-\omega}}{2}$$

or

$$\Gamma(\omega) = -\frac{1}{2\pi} e^{-\omega} + \frac{\pi}{2} - \frac{1}{2\pi} e^{\omega}$$

which is equivalent to the time domain filter

$$-\frac{1}{2\pi} B^{-1} + \frac{\pi}{2} - \frac{1}{2\pi} B$$

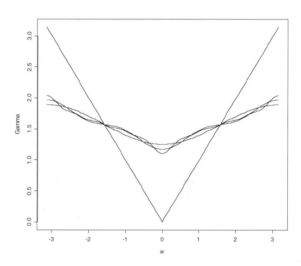

Fig. 6.6 *Filter approximations*

The interested reader who wishes to find out more should consult Bloomfield (1976).

6.3.2 Forecasting

We take the opportunity at the end of this chapter to illustrate the central role of the spectrum in the analysis of stationary time series. Consider the expected mean square error of prediction one step ahead

$$\sigma^2 = E\left[\left(X_{t+1} - \hat{X}_{t+1|t}\right)^2\right]$$

where $\hat{X}_{t+1|t}$ is the optimum predictor at time t. The first question one might ask is under what conditions might σ^2 be zero? We know that if such a perfect prediction is possible then $\{X_t\}$ is a deterministic series. In fact it is possible to go rather further and to show that for a necessary and sufficient condition for a stationary series to be deterministic would be

$$\int_{-\pi}^{\pi} \log(2\pi h(\omega)d\omega) = -\infty$$

Indeed if the integral is bounded then

$$\sigma^2 = \frac{1}{2\pi} \exp\left[\int_{-\pi}^{\pi} \log(2\pi h(\omega))d\omega\right] \qquad (6.10)$$

There is an even more detailed result as follows. If $\int_{-\pi}^{\pi} \log(2\pi h(\omega)) \, d\omega > -\infty$ then $h(w) = \exp(\sigma^2)|A(e^{-i\omega})|^2$ where $A(e^{-i\omega}) = \sum_{j=0}^{\infty} a_j e^{-i\omega j}$ is a one sided Fourier transform. This is equivalent to saying that the time series has an infinite moving average representation. Naturally we have to assume that the cumulative spectrum has a derivative $h(\omega)$. The equation 6.10 has been used as a means of estimating σ^2 by several authors and the problem of finding $A(e^{-i\omega})$ has also been studied by many authors. The prediction problem was solved by Kolmogorov and Wiener at about the same time but in rather different ways. Interested readers should consult Whittle (1963) or Priestley (1981) for further details.

6.4 Aliasing

Sampling a time series introduces an aliasing effect. Suppose we have a series X_t which we observe at every kth time point, say $Y_t = X_{tk}$. Then clearly we must miss some effects. This is certainly true for periodic effects in the series since we have a subtle effect known as aliasing which is caused by sampling.

Suppose we have weekly sales figures and we examine the series every quarter. Then it is clear that we will not observe any monthly effects. Suppose now we sample every seven week weeks but there is a strong four-weekly cycle in the original series. We see the series at weeks 7, 14, 21, ... and the effects we observe originate from weeks which are 3, 2, 1, 4, 3, 2, 1, 4, ... in the cycle. (We divide 7, 14, 21, ... by 4 and then the remainder gives the position in the cycle.) *The upshot is that we observe a cycle of period 4 in our sample series which is induced by the four-weekly effects we cannot see!* This induced effect is known as aliasing and one might expect this to show up in the spectrum. While we have avoided discussion of continuous time series it is quite possible to define autocovariance and spectral functions for such series provided they are stationary. One major difference is that the spectrum of a continuous time series may be defined for all frequencies ω. If we sample such a continuous series we have an aliasing effect.

6.5 Aliasing and the Nyquist frequency

While a time series X_t can often be thought of as having values for all real values of t, for computation we must make recordings at discrete time points. Suppose we choose to digitize the record by taking values at time intervals Δt apart giving $X_{\Delta t}, X_{2\Delta t}, ..., X_{N\Delta t}$ to make inferences about the original series. The choice of time intervals is important since the sampling has two important consequences.

(a) We have no information about phenomena which have frequencies above the Nyquist frequency or folding frequency, of $\frac{1}{2\Delta t}$ cycles per unit time.

(b) These missing effects may distort our perception of those cyclic phenomena which have frequencies below the Nyquist frequency. *This effect is called aliasing.*

Recall that a function $g(t)$ is periodic if

$$g(t) = g(t \pm s) = g(t \pm 2s) = \cdots = g(t \pm ks) = \cdots$$

and the smallest (non-zero) s value is called the period of the function. The frequency f is the number of periods per unit time, that is $f = 1/s$ cycles per unit time. Thus $\cos(2\pi t/s)$ has period s while $\cos(2\pi ft)$ has frequency f. Our Nyquist frequency of $\frac{1}{2\Delta t}$ cycles per unit time will correspond to a minimum period of $2\Delta t$.

As we said before, mathematicians like to work in angular frequencies $\omega = 2\pi f$ radians per unit time but for the moment we look at cycles per unit time for simplicity.

Sampling limits the frequency range because of the nature of periodic functions. Suppose $\Delta t = 1$ and the signal contains a function which is periodic, say $g(t)$ with frequency 3/2 i.e. period 2/3 . Then we observe in sequence

$$g(0), g(1/3), g(2/3) \ldots = g(0), g(1/3), g(2/3), \ldots$$

a signal which repeats in steps of 3 giving an apparent frequency of 1/3. If we regard our periodic component $g(t)$, with period $f_b = k/2\Delta t + f_0, f_0 < f_N$, as being expressed in terms of complex exponentials $\exp(i2\pi f_b t)$ then, since $t = m\Delta t$ they become $\exp(i2\pi f_b m \Delta t) = \exp(i2\pi f_0)$. For an example see Figure 6.7 where we see that if we make observations at integer times we see identical values from two cosines even though they have different frequencies, since the frequencies exceed the Nyquist and so are aliased.

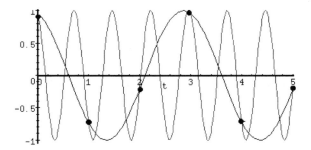

Fig. 6.7 *Example of aliasing*

If the power spectrum of the original series is $h(f)$ then we can show that the power spectrum of the observed, digitised, series $h_d(f)$ is

$$h_d(f) = \sum_{-\infty}^{\infty} h(j + k/\Delta t) \qquad -\frac{1}{2\Delta t} \le f \le \frac{1}{2\Delta t}$$

where h is the spectrum of the original continuous series. Our observed spectrum is thus the result of folding the original over the Nyquist range. This means that the observed value of the power spectrum at f_0 is made up not only of $h(f_0)$ but also of the values of the original spectrum at the aliases to f_0, that is $f_0 \pm 1\Delta t$, $f_0 \pm 1/2\Delta t$, $f_0 \pm 1/3\Delta t$, ... So when you digitize a sequence it is vital that there are no components with appreciable power whose frequency lie outside the Nyquist range. Indeed we can prove that if the power outside this range is exactly zero then the original series can be exactly reconstructed from the digitized one! This of course presupposes a continuous time series which has a spectrum. More often we sample a discrete time series, say we observe Y_t where $Y_t = X_{tr}$ for some constant r. In this case we get a similar situation. Neave (1970) showed that the spectrum of Y_t , say $h_Y(\omega)$ can be written in terms of the X_t spectrum $h_Y(\omega)$ as follows

$$h_Y(\omega r) = \frac{1}{r}\alpha(r, \omega/r)$$

and

$$\alpha(r, \omega) = h_X(\omega) + h_X\left(\frac{2\pi}{r} - \omega\right) + h_X\left(\frac{2\pi}{r} - \omega\right) + \cdots + h_X\left(\frac{2[r/2]\pi}{r} - \omega\right)$$

So if r is 2

$$h_Y(\omega r) = \frac{1}{2}(h_X(\omega/2) + h_X(\pi - \omega/2) + h_X(\pi + \omega/2))$$

A comprehensive account is given in Koopmans (1974) and Priestley (1981).

6.6 Exercises

1. Suppose the series X_t is defined by $X_t = \sum_{k=1}^{n} A_k \cos(\omega_k t + \theta_k)$ where n; A_k, $\omega_k k = 1, 2, \ldots, n$ are known constants and the $\theta_k k = 1, 2, \ldots, n$ are independent random variables having a uniform distribution on $[-\pi, \pi]$. Assuming that

$$\rho(s) = \frac{\sum_{k=1}^{n}(A_k/2)^2 \cos(\omega_k s)}{\sum_{k=1}^{n}(A_k/2)^2} \qquad s = \ldots, -1, 0, 1, 2, \ldots$$

   show that the cumulative spectrum $F(\omega)$ is a step function with jumps at ω_k $k = 1, 2, \ldots, n$. What are the magnitude of the jumps? You may find it worth setting $n = 1$ if you find the algebra too complex.

2. The spectrum of X_t is known to be $F(\omega) = \begin{pmatrix} 0 & \omega < -\pi/6 \\ 1 & -\pi/6 < \omega/6 \\ 2 & \pi/6 \le \omega < \pi \\ 2 & \omega = \pi \end{pmatrix}$

   Suggest a model for X_t.

3. Assuming that X_t is white noise find the spectrum of $Y_t = e^{i\theta t} X_t$.

4. Suppose the stationary series X_t has a spectrum $h(\omega)$. Find the spectra of (i) $(1 - B)X_t$ (ii) $(1 - B^{12})X_t$ in terms of $h(\omega)$.

5. Suppose $Z_t = X_t + Y_t$ where X_t and Y_t are stationary and uncorrelated with spectra $h_x(\omega)$ and $h_y(\omega)$. Find the spectrum of Z_t in terms of $h_x(\omega)$ and $h_y(\omega)$.

6. Suppose that $X_t = \alpha X_{t-1} + \epsilon_t$ where ϵ_t is zero mean white noise. If $Y_t = \beta Y_{t-1} - X_t$ find the spectrum of X_t. Find the spectrum of Y_t when

   (i) $\alpha = 0$
   (ii) $\alpha = \beta$
   (iii) $\alpha \ne \beta$ $\alpha \ne 0$ and $\beta \ne 0$
   (iv) $\alpha \ne 1$ and $\beta = 1$

7. Suppose $Y_t = \frac{1}{P}\sum_{j=0}^{P-1} X_{t-Qj}$ where X_t is a stationary zero mean process and P,Q are constants. Calculate the transfer function of the filter and sketch the gain when P=4 and Q=6. An economist has a time series with a strong 3-monthly cycle. How should one use the above filter to remove such a cycle?

8. X_t is stationary with a zero mean. Define Y_t as $Y_t - x_{2t}$. Show that the spectral density of Y_t is given by

$$h_y(\omega) = \frac{1}{2}[h_x(\omega/2) + h_x(\omega/2 - \pi)]$$

   When $X_t = \theta X_{t-1} + \epsilon_t$, where ϵ_t is zero mean white noise, find the spectral density of X_t. Hence obtain the spectral density of Y_t and from it deduce a model for Y_t. Check your solution by evaluating the Y_t model in the time domain.

9. Sketch the power spectra for the following series

   (i) A monthly series with a strong cycle of period 12 months.
   (ii) A series which is almost white noise.
   (iii) A white noise series which was differenced by mistake.
   (iv) An AR(1) model with a large and positive parameter
   (v) An AR(1) model with a large and negative parameter

10. The spectrum of an economic series shows undesirable peaks at the 'seasonal' frequencies

$$f = 1/12, 2/12, 3/12, 4/12, 5/12, 6/12$$

cycles per year using monthly data. Suppose we consider filtering the data with one of the following filters

A : $Y_t = (1 - B)X_t$
B : $Y_t = (1 - B^{12})X_t$
C : $Y_t = (1 - B)U_t$ where $U_t = (1 - B^{12})X_t$

Find the gain and phase of each filter and describe in words the effect of each on the series. Which of the filters could be described as a high-pass filter and why? Which filter(s) eliminate the seasonal frequencies? Under what conditions might C be superior to B?

7
Estimation and use of the power spectrum

Since the spectrum is

$$h(\omega) = \frac{1}{2\pi} \sum_{s=-\infty}^{\infty} \gamma(s) \cos(sw)$$

the obvious approach is to replace the autocorrelations $\gamma(s)$ by their estimates giving

$$\hat{h}_p(w) = \frac{1}{2\pi} \sum_{s=-m}^{m} \hat{\gamma}(s) \cos(s\omega)$$

where m is some suitable number of autocorrelations. Depending on our choice of covariance estimates we find that $2\pi\hat{h}_p(\omega)$ is equal to very close to

$$\frac{1}{N} \left| \sum_{t=1}^{N} X_t e^{-i\omega t} \right|^2$$

The expression

$$I(\omega) = \frac{1}{N} \left| \sum_{t=1}^{N} X_t e^{-i\omega t} \right|^2$$

is known as the periodogram and is closely related to the discrete fourier transform (DFT) of the series

$$J(\omega) = \frac{1}{\sqrt{N}} \sum_{t=1}^{N} X_t e^{-i\omega t}$$

Unfortunately the periodogram is well known to be a poor estimate of the power spectrum. While it has many useful properties it is not a consistent estimator and is an erratic fluctuating function. Despite the variability the periodogram has rather nice distributional properties.

7.1 Properties of the periodogram

If X_t is stationary with mean μ and with an absolutely convergent covariance function then

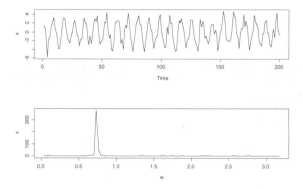

Fig. 7.1 *Periodogram of* $3\cos(0.7t) + $ *noise and the original trace*

1. $E[I(\omega)] = 2\pi h(\omega)$ $\omega \neq 0$ and $E[I(\omega)] = 2\pi h(\omega) + n\mu^2$ $\omega = 0$
2. $I(\omega, x) \simeq 2\pi h(\omega) I(\omega, \epsilon)$. where $I(\omega, x)$ is the periodogram based on X_t while $I(\omega, \epsilon)$ is the periodogram based on a white noise series.
3. $I(\omega_1), I(\omega_2), \ldots, I(\omega_k)$ are independent for distinct Fourier frequencies $\omega_1, \omega_2, \ldots, \omega_k$.
4. $I(\omega)/\pi h(\omega)$ is distributed as Chi-squared with two degrees of freedom when $\omega \neq 0, \pi$ and as Chi-squared with one degree of freedom at the end points 0 and π.

Perhaps the main reason for initial interest in the periodogram was its value in the detection of periodicities. Recall that the spectrum of a series containing a sinusoid with period p will have a discontinuity at the corresponding frequency. We thus expect a jump in the periodogram at this frequency. This is easily seen by generating an artificial series. Figure 7.1 shows that the periodogram has a spike at around 0.7 radians. This implies a cyclic series ith a period of 7.

An even more dramatic example is the periodogram of the ECG trace shown in Figure 7.2 where it is quite clear that there are two important periodic effects in the data at $\omega = 1.25$ and at $\omega = 2.51$.

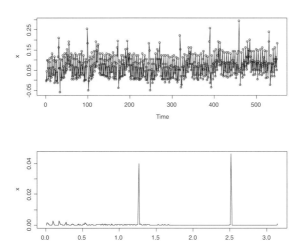

Fig. 7.2 *Periodogram of an ECG trace together with the original trace*

If we look at the sunspot series

```
Wolfer sunspot series 100 observations
101     82     66     35     31      7     20     92    154    125
 85     68     38     23     10     24     83    132    131    118
 90     67     60     47     41     21     16      6      4      7
 14     34     45     43     48     42     28     10      8      2
  0      1      5     12     14     35     46     41     30     24
 16      7      4      2      8     17     36     50     62     67
 71     48     28      8     13     57    122    138    103     86
 63     37     24     11     15     40     62     98    124     96
 66     64     54     39     21      7      4     23     55     94
 96     77     59     44     47     30     16      7     37     74
```

we can see from the plot, in Figure 7.3, that the periodogram has a peak at around frequency 0.7 that is lag 9. Note the mean was removed!

The periodogram can be, as we have said, an erratic and peaky function and we may have problems in deciding if any given peak is real. A test for just this problem was devised by Fisher and works as follows. Assume that the series is noise plus a mean level and suppose for simplicity we exclude frequencies 0 and π and let

$$E_k = I(\omega_k)/(\pi h(\omega_k)) \qquad k = 1, 2 \dots m$$

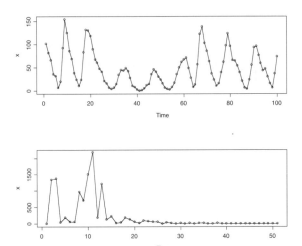

Fig. 7.3 *Sunspot series and periodogram*

for some suitable value m. Since the E_k are independent, each having a chi-squared distribution, then we can find the distribution of the sum. If one of the values, say E_j is the largest peak then we compute

$$M_j = E_j \bigg/ \sum_{k=1}^{m} E_k$$

Fisher (1929) showed that

$$p(M_j > x) = \sum_{j=1}^{k} (-1)^{j-1} \binom{m}{j} (1 - jx)^{m-1}$$

For details refer to Wilks (1962). If we assume that the spectrum is flat for the sunspots the maximum is 13 663.624 at lag 11, corresponding to an angular frequency of 0.705 and the sum of the non-zero frequencies is 67 768.9 giving $M_k = 0.20$. Then $p[M_k] > 0.20 = 0.001$ indicating a seasonal effect. This conclusion is not quite as clear cut as you might think since obviously our test assumes for the null hypothesis that the series is noise. In the case of the sunspots we do not know if this is true (it is not) and so we need to modify our procedure.

7.2 The Discrete Fourier Transform (DFT)

The Discrete Fourier Transform is fairly obviously like a truncated Fourier transform of the data series. In fact it is a proper transform in the sense that if we know the DFT we can reconstruct the series using the inverse transform. Suppose we have

$$J(\omega_j) = \frac{1}{\sqrt{N}} \sum_{t=0}^{N-1} X_t e^{-i\omega_j t} \text{ for } \omega_j = 2/\pi j/Nj = 0, 1, 2, \ldots, N-1$$

then

$$X_t = \frac{1}{\sqrt{N}} \sum_{j=0}^{N-1} J(\omega_j) e^{i\omega t}$$

The real motivation for studying the DFT is the usual one for looking at any generating function – there are some operations which are easily done using generating functions and which are rather difficult otherwise. In time series it is the study of convolutions which makes the generating function so valuable.

Suppose we have X_t and Y_t then the convolution is

$$Z_t = \sum_{u=0}^{N-1} X_u Y_{t-u} \qquad t = 0, 1, \ldots, N-1$$

This is an unpleasant calculation but the DFT of Z is just the product of DFTs of the X series! While DFTs have many interesting properties we leave the reader to pursue them elsewhere.

We will expand on just one application which is to smoothing.

7.3 Smoothing by DFT

If we look at a DFT we see that a series is represented by a finite number of frequency terms. What we can do is set all the frequencies above a fixed value, say ω_0 to zero. If we now reconstruct the series from the modified DFT we find that the parts of the series corresponding to the high frequency terms are lost and we have a smoother series. We can of course do the reverse and set the low frequencies to zero. In this case the resulting reconstructed series is left with the more variable terms. The choice of the cut off frequency is ω_0 and is not known if we have no a priori information. It is however an explicit value which we are forced to set and in consequence determine the smoothness we require.

Figure 7.4 shows the display of an ECG series (mean removed) and the smoothed version of the series where only the bottom 10% of the frequencies are retained. As one can see there is considerable smoothing.

A more dramatic example is given in Figure 7.5. Here the series has been dramatically smoothed (0.1% of the frequency range remains) to bring out the longer period effects.

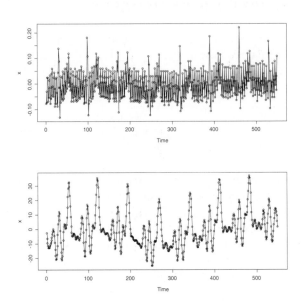

Fig. 7.4 *An ECG series and a smoothed version*

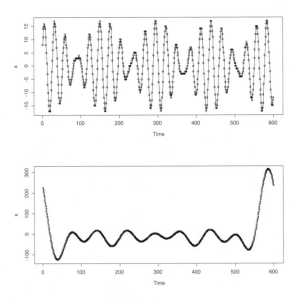

Fig. 7.5 *Magnitude of a star and the same series smoothed*

7.4   The Fast Fourier Transform

Suppose we have a series of length N. To compute the periodogram at some N frequencies we need to evaluate the DFT

$$J(\omega) = \frac{1}{\sqrt{N}} \sum_{t=1}^{N} X_t e^{-i\omega t}$$

N times and each DFT computation involves N terms each of which requires a calculation. Until the 1960s this was a challenge to most computing systems, indeed for large N values such as are found in geophysical exploration for example, the computations can still be a challenge. It is easy to make some simple savings in calculation but the real timesaver was the Fast Fourier Transform (FFT) which was (re)discovered in an important paper by Cooley and Tukey (1965). The FFT reduced the computational time required to evaluate DFTs dramatically and as a consequence it became natural to base calculations on the DFT. So for example one might compute the autocorrelations by first estimating the spectrum and then obtaining the autocorrelations by using the fact that they are Fourier coefficients.

FFT programmes are common and can be found in most numerical software suites although it is often a challenge to determine quite what they do. Thus in R for example we have

```
Fast Discrete Fourier Transform

     fft(z, inverse = FALSE)
     mvfft(z, inverse = FALSE)


Arguments:

     z: a real or complex array containing the values to
        be transformed

  inverse: if 'TRUE', the unnormalized inverse transform is
           computed (the inverse has a '+' in the exponent of
           e, but here, we do not divide by '1/length(x)').

Value:

     When 'z' is a vector, the value computed and returned
     by 'fft' is the unnormalized univariate Fourier
     transform of the sequence of values in 'z'.  When 'z'
     contains an array, 'fft' computes and returns the mul-
     tivariate (spatial) transform.  If 'inverse' is 'TRUE',
     the (unnormalized) inverse Fourier transform is
     returned, i.e., if 'y <- fft(z)', then 'z' is 'fft(y,
     inv=TRUE) / length(y)'.
```

By contrast, 'mvfft' takes a real or complex matrix as
argument, and returns a similar shaped matrix, but with
each column replaced by its discrete Fourier transform.
This is useful for analyzing vector-valued series.

The FFT is fastest when the length of the series
being transformed is highly composite (i.e. has many
factors). If this is not the case, the transform may
take a long time to compute and will use a large amount
of memory.

References:

Singleton, R. C. (1979). Mixed Radix Fast Fourier
Transforms, in Programs for Digital Signal Processing,
IEEE Digital Signal Processing Committee eds. IEEE
Press.

While we will just use the FFT as a black box it is interesting to see the basic
idea.
Define a scaled DFT as above

$$D(\omega) = \sum_{t=1}^{N} X_t e^{-i\omega t}$$

Assume that N is a power of two, then we can define

$$D_1(\omega) = \sum_{t\text{even}}^{N} X_t e^{-i\omega t}$$

and

$$D_2(\omega) = \sum_{t\text{odd}}^{N} X_t e^{-i\omega t}$$

Then

$$D(\omega) = D_1(\omega) + D_2(\omega) = \sum_{t=0}^{(N/2)-1} X_{2t} e^{-i\omega 2t} + \sum_{t=0}^{(N/2)-1} X_{2t+1} e^{-i\omega(2t+1)}$$

After some algebra this can be written as

$$D(\omega) = D_1(\omega) + \exp(-2\omega) D_2(\omega)$$

The implication is that the DFT is the weighted sum of two shorter DFTs. If N
is a power of two then we can replace the shorter DFTs by shorter ones and so

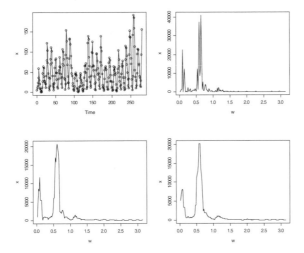

Fig. 7.6 *Long sunspot series with its scaled periodogram and two smoothed periodograms*

on. This means that the original DFT can be written as the sum of order 1 DFTs and some 'twiddle factors'. Algorithms using this idea are very fast and stable. Of course not all series are powers of two but the idea can be extended to give savings by decomposing N into prime powers. There is a drop in speed but savings can still be made. For more detail see Camina and Janacek (1984).

7.5 Estimating the spectrum: nonparametric estimates

As we have seen, the periodogram, suitably scaled, is not a very usable spectral estimate. One possibility is to modify the periodogram so as to remove the drawbacks we have identified. This approach, we shall refer to as the non-parametric or windowed approach, is well established. As the periodogram fluctuates wildly one possibility is to smooth the function so as to make it more tractable. A simple averaging approach works reasonably well. In Figure 7.6 we show the long sun spot series, the periodogram of the series and two smoothed versions, with simple averaging of 3 and 5 points. As we see the simple average does smooth the periodogram quite markedly.

A more refined possible estimate is $\hat{h}(\omega)$ where

$$\hat{h}(\omega) = \frac{1}{2\pi} \int_{-\pi}^{\pi} W(\omega - \theta) I(\theta) \, d\theta$$

where $W(\theta)$ is a suitably chosen function, which decays to zero as $|\theta|$ increases and integrates to one. In fact since the periodogram is calculated at discrete frequency points we should really have a sum

$$\hat{h}(\omega) = \frac{1}{2\pi N} \sum_{j=1}^{N} W(\omega - \omega_j) I(\omega_j)$$

The idea being to average the periodogram ordinates near the frequency of interest ω. It is rather convenient however to use the integral form as a notation. The problem now is to select the window function $W(\theta)$ so as to ensure a reasonable estimate. For a sharply peaked function we can approximate crudely as follows

1.

$$E[\hat{h}(\omega)] = h(\omega) \int_{-\pi}^{\pi} W(\theta) \, d\theta$$

2.

$$var(\hat{h}(\omega)) = \frac{2\pi}{N} \int_{-\pi}^{\pi} W^2(\omega - \theta) h(\theta)^2 \, d\theta$$

3.

$$cov(\hat{h}(\omega)\hat{h}(\psi)) = \frac{2\pi}{N} \int_{-\pi}^{\pi} W(\omega - \theta) W(\psi - \theta) h(\theta)^2 \, d\theta$$

We shall simply ignore any correction required for the end frequencies $0, \pi$. We see that the variance of the estimate depends on the shape of the function W. These results can be sharpened quite considerably (see for example Priestley (1981)). There have been many suggestions as to the form of this function of which we give only two.

1. The Parzen or Frejer Window

$$W(\omega) = \frac{3}{8\pi M^3} \left[\frac{\sin(M\omega/4)}{2\sin(\omega/4)} \right]^4 \{1 - (2/3)[\sin(\omega/2)]^2\}$$

2. The Bartlett Window

$$W(\omega) = \frac{1}{2\pi M} \left[\frac{\sin(M\omega/2)}{\sin(\omega/2)} \right]^2$$

Table 7.1 *Properties of spectral windows*

Window	Bandwidth	Variance/$h(\omega)^2$	EDF	Bias
Unit	$2\frac{\pi}{M}$	$2.00\frac{M}{N}$	$\frac{N}{M}$	no simple form
Bartlett	$2\frac{\pi}{M}$	$\frac{2M}{3N}$	$\frac{3N}{M}$	$-\frac{1}{M}h^{[1]}(\omega)$
Parzen	$\frac{8\pi}{3M}$	$0.54\frac{M}{N}$	$3.7\frac{N}{M}$	$\frac{6}{N^2}h''(\omega)$

We use the most complex form of the Parzen window – many texts omit the $\{1 - 2/3[\sin(\omega/2)]^2\}$ term. There has been much argument about the choice of windows but in fact the effects of using differing windows are very small. They do give rise to slightly different estimates but for our purposes we shall ignore these second order effects.

Concentrating the window around the frequency of interest gives the greatest weight to these frequencies which is generally a good idea but if the window has subsidiary peaks, so called 'side lobes' then the estimate at ω may be contaminated by effects at other frequencies. The resulting distortion is called leakage. Differing windows control leakage in different ways but the windows that we suggest are satisfactory.

To get some feel for the parameters we require in order to choose the right form of window we must define the peakedness or bandwidth of the window function. There are many definitions of bandwidth, all slightly different but all giving much the same answer. We define the bandwidth Bw as *the width of a rectangular window having the same maximum height as $W(\omega)$ and the same area in the frequency of interest.* Thus $B_w = \frac{1}{W(0)}\int_{-\pi}^{\pi} W(\theta)\,d\theta$ which since the area of the window is units gives $B_w = \frac{1}{W(0)}$

The distribution of the weighted estimate is clearly a sum of chi-squared variables and an approximation to the distribution is the chi-squared distribution itself. Since for chi-squared the degrees of freedom is $2(\mu/\sigma)^2$ we define the equivalent degrees of freedom in the same way. We take three of the many possible windows, The Unit, Bartlett, and Parzen where the Unit is

$$W(\omega) = 1 \text{ for } -\left[\frac{(M-1)}{2}\right] \leq \omega \leq \left[\frac{M}{2}\right]$$

These have differing bandwidths etc. as can be seen from Table 7.1.

Notice here $h^{[1]}(\omega) = \frac{1}{2\pi}\sum_k |k|\gamma(k)\exp(-i\omega k)$

Bandwidth, variance, etc. cannot alone be criteria for the choice of window. The shape of the window is also important if the average we choose is to give a well resolved estimate. By this we mean it does not change very much over the bandwidth of the window. We tend to choose to use the Parzen window, primarily because we find it reasonable and it must give non-zero estimates.

The choice of M is more problematic. Obviously we would like to make M as large as possible so as to decrease the bandwidth. If we do so then the variance of the estimate at any frequency must increase. Thus we need to find some compromise value for M.

The usual pragmatic approach is to try values of M between N/3 and N/5. As M increases the estimate becomes smoother and we choose a value that seems 'smooth enough'. This is rather vague but it essentially means that there is a peak in the spectrum which has the minimum bandwidth of interest, say B, that we wish to resolve. If we specify this then we can choose the bandwidth of the smoothing kernel, say B_w, to fit this desired resolution. We require $B_w \leq B_f$ and we take as a reasonable choice $B_w \leq B_f/2$.

Without a minimum bandwidth B_f any choice of the parameter M is somewhat arbitrary. It is important to realize that for a fixed length of realization the requirements of Bandwidth and Variance are contradictory. If one decreases the other increases, and since the amount of information contained in any data segment is finite this is only reasonable and we need to come to some sensible compromise. This is exactly the same as the problems of statistical testing where one has two error probabilities. In the optimal case one chooses the length of realization to achieve the required bandwidth and variance. This may not always be possible in which case compromise is essential.

7.5.1 Lag windows

We can look at these smoothed estimates in a rather different light if we assume that $W(\omega) = \frac{1}{2\pi} \sum_{j=-\infty}^{\infty} \lambda_j e^{-i\omega j}$ for in this case

$$\hat{h}(\omega) = \frac{1}{2\pi} \sum_{j=-M}^{M} \lambda_j \hat{\gamma}(j) \exp^{i\omega j} \tag{7.1}$$

for some parameter $M \leq N$. As we see this is a weighted sum of estimated autocorrelations, the weight sequence $\{\lambda_j\}$ being known as the lag window. For the windows we have discussed so far the corresponding lag windows are given in Table 7.2. We can in consequence think of the smoothed periodogram as a weighted sum of covariances, the trick being to choose a suitable sequence $\{\lambda_j\}$ or window $W(\omega)$. This lag window approach has fallen into disfavour, the reason being a computational one. The periodogram is essentially a DFT and in consequence can be computed very quickly using an FFT. This being so, the obvious approach is to compute the periodogram and then to smooth it. Nevertheless the lag window approach does have some virtues, especially for short data lengths. Indeed we have often used this approach to compute spectra, even on a desktop machine. If one stores the sines and cosines it can prove to be quite efficient for shortish ($N < 300$) data lengths. The computations are much easier and this may be important if one is using a macro in a package like MINITAB. As we said above since

$$J(\omega) = \frac{1}{\sqrt{N}} \sum_{t=1}^{N} x_t e^{-i\omega t}$$

Table 7.2 *Lag window properties*

Window	
	1 for $\|s\| \leq M$ 0 otherwise
Unit	$1 - \|s\|/M$ for $\|s\| \leq M$ 0 otherwise
Bartlett	$1 - \|s\|/M$ for $\|s\| \leq M$ 0 otherwise
Parzen	$1 - 6(\|s\|/M)^2 + 6(\|s\|/M)^3$ for $\|s\| \leq M/2$ $2(1 - \|s\|/M)^3$ for $M/2 \leq \|s\| \leq M$ 0 otherwise

is essentially a DFT it makes sense to compute the transformation and hence the periodogram using an FFT. Given the periodogram then a Daniell, or similar window becomes attractive since we work directly with $W(\omega)$. The correlations can also be computed using an FFT since the autocovariances are just the Fourier transforms of the spectrum. Most available programmes use the FFT approach, since it can be considerably faster. There are some practical problems especially when N is not a nice product of powers of primes. In this case the computational advantage of the FFT is smaller. One possibility is to 'pad' the series with zeros in order to make N a nice number viz.

$$Y_t = X_t \qquad t \leq N \tag{7.2}$$
$$Y_t = 0 \qquad t = N+1, N+2, \ldots, 2^n. \tag{7.3}$$

This can be taken a little further and results in the tapering of a series. Suppose we define a new series $\{Y_t\}$ from the observed $\{Y_t\}$ by $\{Y_t = a_t X_t\}$ where the coefficients a_t are from a decaying sequence of constants.

Teukolsky *et al.* (1986, pp. 423–4), note that 'when we select a run of N sampled points for periodogram spectral estimation, we are in effect multiplying an infinite run of ... data ... by a window function in time, one which is zero except during the total sampling time, and is unity during that time.' The sharp edges of this window function contain much power at highest frequencies, which is imparted to the windowed signal and leads to power leakage. Weighting the data (or correlation function) is an accepted traditional approach to reducing power leakage and one we will expand upon later.

7.5.2 Some examples of non-parametric spectral estimates

We have looked at the global mean temperature series before (Figure 7.7) so we will now look at the spectrum of the series. As we see from Figure 7.8, the spectrum of the differenced series has little power at low frequencies suggesting

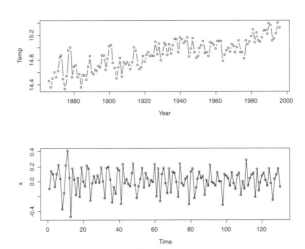

Fig. 7.7 *Spectrum of differenced GMT series, using averages of 5 on the periodogram*

that there are not any significant long range effects in the differences. The trend in the series is thus removed by differencing suggesting that it is constant or at most linear in time. The peak at around 1.6 radians, 32 years is interesting but it is not a true cyclic effect. There is also considerable power at high frequencies so we might expect short term fluctuations. In fact if we use a window of the Parzen type we see that the height frequencies are an artifact of the smoothing and there is not a great deal of randomness in the series. We suggest that the reader use a sensible window, e.g. Parzen for spectral estimation!

The Parzen window based spectrum (Figure 7.10) of the Beveridge wheat price series has no really interesting cycles to reveal. There are small high frequency (short term) effects so the series is stable and slowly changing and there is a suggestion of a concentration of variance around a frequency of 0.4 radians/year, a period of around 15 years. This is not however a real cycle!

7.5.3 Sampling properties of the smoothed spectral estimate

Given that the estimates we have considered are weighted sums of periodograms and the periodograms are independent chi-squared variables we would expect to be able to approximate the distribution of our spectral estimates by a chi-squared distribution. It can be shown that the chi-squared approximation is a reasonable one if not quite correct and we can summarize the result as follows.

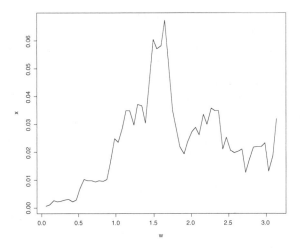

Fig. 7.8 *GMT series and differences*

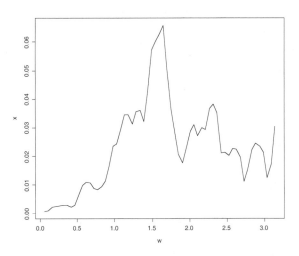

Fig. 7.9 *Spectrum of differenced GMT series, using Parzen window*

Table 7.3 *Beveridge annual wheat price index 1500–1711, read by rows*

106	118	124	94	82	88	87	88	88	68	98	115	135	104	96	110
107	97	75	86	111	125	78	86	102	71	81	129	130	129	125	139
97	90	76	102	100	73	86	74	74	76	80	96	112	144	80	54
69	100	103	129	100	90	100	123	156	71	71	81	84	97	105	90
78	112	100	86	77	80	93	112	131	158	113	89	87	87	79	90
90	87	83	85	76	110	161	97	84	106	111	97	108	100	119	131
143	138	112	99	97	80	90	90	80	77	81	98	115	94	93	100
99	100	94	88	92	100	82	73	81	99	124	106	106	121	105	84
97	109	148	114	108	97	92	97	98	105	97	93	99	99	107	106
96	82	88	116	122	134	119	136	102	72	63	76	75	77	103	104
120	167	126	108	91	85	73	74	80	74	78	83	84	106	134	122
102	107	115	113	104	92	84	86	101	74	75	66	62	76	79	97
134	169	111	109	111	128	163	137	99	85	72	88	77	66	64	69
125	175	108	103												

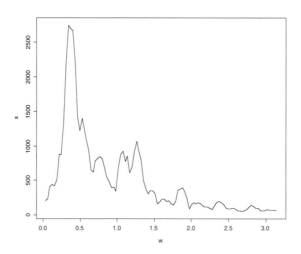

Fig. 7.10 *Spectrum of Beveridge wheat price index*

Distributional properties

1. The spectral estimate $\hat{h}(\omega)$ has a distribution which is approximately chi-squared with n degrees of freedom. The equivalent degrees of freedom (EDF) ν is defined as

$$\nu = 2\frac{\left(E[\hat{h}(\omega)]\right)^2}{var(\hat{h}(\omega))}$$

   since for a chi-squared variable with n degrees of freedom the mean is ν and the variance 2ν. As can be seen from the table the higher the degrees of freedom the smaller the variance and the greater the smoothing of the periodogram.

2. Spectral estimates at least one bandwidth apart will be assumed to be independent.

Given this approximate distribution it is easy to produce a confidence interval for the $\hat{h}(\omega)$. If we are given α then we select the $\alpha/2$ quantiles of the chi-squared distribution with ν degrees of freedom say $p[\chi^2 \leq a] = p[\chi^2 \geq b] = \alpha/2$ then

$$p\left[a \leq \nu\frac{\hat{h}(\omega)}{h(\omega)} \leq b\right] = 1 - \alpha$$

so the $(1 - \alpha)100\%$ confidence interval is $\nu\frac{\hat{h}(\omega)}{b}$, $\nu\frac{\hat{h}(\omega)}{a}$. This gives a pointwise estimate rather than a confidence interval over a frequency band. For most cases this will suffice. One can find a band over all frequencies and in this case the reader is referred to Priestley (1981, Chapter 6). An example of a chi squared pointwise interval is given in Figure 7.11 where confidence intervals are supplied for the Beveridge wheat price series spectrum. The degrees of freedom for a Parzen window are $3.7N/M$. In our case using a truncation point of a quarter of the series, i.e. $M = N/4$ $\nu = 14$. The consequent confidence 95% intervals give some idea of the variation in the spectrum but should be used with some care when the spectrum is rapidly changing.

Some authors have used central limit theorem arguments to use a normal approximation for the distribution of $\hat{h}(\omega)$. In the limit one can prove that $\hat{h}(\omega_1)$, $\hat{h}(\omega_2), \ldots, \hat{h}(\omega_k)$ have a joint normal distribution with variances and covariances of the forms shown at the beginning of this chapter. In this case the $(1 - \alpha)100\%$ confidence interval for $h(\omega)$ is

$$\hat{h}(\omega)\left(1 \pm z_{1-\alpha/2}\sqrt{\frac{2}{\nu}}\right)$$

where $z_{1-\alpha/2}$ is the $1-\alpha/2$ quantile of the standard normal distribution. The problem of these expressions is that they depend on the value of $\hat{h}(\omega)$ at frequency ω.

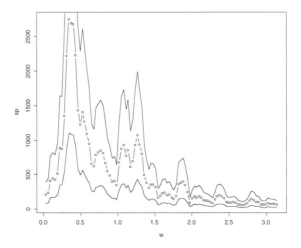

Fig. 7.11 *Beveridge spectrum with confidence intervals*

It is rather more satisfactory to consider $log\hat{h}(\omega)$. The corresponding confidence intervals are

$$\log(\hat{h}(\omega)) + \log(\nu/b), \log(\hat{h}(\omega)) + \log(\nu/a)$$

have uniform width and are much easier to handle. There are advantages to the plotting of the log spectrum and this is commonly done by engineers.

7.5.4 Tapering and prewhitening

Two common techniques used for cutting down the bias in periodogram based estimates are prewhitening and tapering. Prewhitening is simple, at least in concept, as the most difficult spectra to estimate are those with sharp peaks where the aim is to flatten the spectrum by filtering. In practice it is rather more complex since to design a filter to perform this 'whitening' we need to know the form of the function we wish to estimate. Nevertheless it can be a valuable option and it is common for an approximate AR to be fitted to the data and for this to be used as a prewhitening filter.

Tapering is another bias reduction technique which we have mentioned before, in which the data are adjusted by multiplying the data series by a sequence of constants h_t. The resulting values $y_t = h_t x_t$ are then used as raw input for the spectral analysis. If we regard the time series as extending into the infinite future then our finite sample $x_1, \ldots, x_N$ is a tapered version of the infinite series with

$h_t = 0$ for t exceeding N and $h_t = 1$ for $t \le N$. The periodogram is then

$$\frac{1}{N} \left| \sum_{t=1}^{N} h_t x_t e^{i\omega t} \right|^2$$

and we can show that the smoothed spectral estimate based on the tapered data has an effective smoothing window of the form where and $A(\omega)$ is the Fourier Transform of the taper sequence. The 'finite taper' above has a rectangular shape whose sharp corner gives a pronounced ringing in $A(\omega)$, caused by the Gibbs phenomenon. The taper should smooth this corner. An important and interesting extension of this idea called multitapering was proposed by Thomson (1982) and is discussed below. Examples of tapers are the trapezium taper which is 1 except towards the end of the data series when it decays linearly to zero, another is the cosine taper

$$h_t = [1 - \cos(2\pi t/N)] \qquad t = 1, 2, \ldots, N$$

which is sometimes known as the Hanning taper. As always one pays a price for using a taper, and in the price is an increase in variance. This is hardly surprising since we are in essence reducing the data length. Priestley (1981) shows that if h_t can be written as $h_t = a(t/N)$ then the asymptotic variance of the spectral estimate is multiplied by

$$\left(\int_0^1 a^4(u)\, du \right) \Big/ \left(\int_0^1 a^2(u)\, du \right)^2$$

Our hope is of course that the inflation in the variance is accompanied by a corresponding reduction in the bias.

7.6 Multitaper estimates

An interesting development in non-parametric spectral estimation has been the multitaper estimates proposed by Thomson (1982). This is an attempt to keep the advantages of tapering, and remove bias by controlling leakage, with a reduced variance. The methodology is quite simple. Given a series we choose a set of K tapers, say $\{h_{j,k}, j = 1, 2, \ldots, N; j = 1, 2, \ldots, K\}$. For every taper we compute the sample spectrum, so for the kth taper we have the

$$\hat{h}_k(\omega) = \frac{1}{2\pi N} \left| \sum_{t=1}^{N} h_{t,k} x_t e^{-i2\pi t \omega} \right|^2$$

The multitaper spectral estimate is then just the average of these 'eigenspectra', that is

$$\hat{h}_m(\omega) = \frac{1}{K} \sum_{k=1}^{K} \hat{h}_k(\omega)$$

If we choose the tapers to be orthonormal, that is

$$\sum_{t=1}^{N} h_{t,j} h_{t,k} = 0 \text{ if } j \neq k \text{ and } = 1 \text{ if } j = k$$

then the eigenspectra are approximately uncorrelated and we can show that $2K\hat{h}_m(\omega)/h(\omega)$ is distributed as chi-squared with $2K$ degrees of freedom. So we retain the frequency resolution of the simple tapered estimate and gain degrees of freedom.

A particularly good set of tapers (Percival and Waldren (1993)) are the discrete prolate spheriodal sequences (DPSS) but they do involve a computational overhead. We shall restrict ourselves to the sine tapers introduced by Riedel and Sidorenko (1993)

$$h_{t,k} = \sqrt{\frac{2}{N+1}} \sin\left(\frac{(k+1)\pi t}{N+1}\right) \qquad t = 1, 2, \ldots, N$$

It is not difficult to show that

1. $E[\log(\hat{h}_m(\omega)/h(\omega))] = \psi(K) - \log(K)$
2. $var[\log(\hat{h}_m(\omega)/h(\omega))] = \psi'(K)$

where ψ and ψ' are the digamma and trigamma functions.

It is simple to write programs for multitaper spectra, at least if we use sine tapers, the question being the choice of the number of tapers to use. It can be shown, and here we suggest that the reader refer to Percival and Waldren (1993), that resolution falls off as K increases so again we have a trade off between variance and resolution. We suggest K of the order of 3 and that some experimentation is sensible. The Beveridge multitaper spectra are shown in Figure 7.12, the top version being K=3 while the lower is K=2. There is a perceptible decrease in sharpness as K increases! The author finds that the multitaper spectra have much to commend them, they work well and have simple chi-squared distributions so approximate confidence intervals are no problem, they are simple to use and the programs (at least ones we provide in R) are fast and efficient.

As a further example we show in Figure 7.13 the multitaper spectrum of the lynx series (scaled by the standard deviation). Here K=3 and the main interest is the power bulge between 0.5 and 1 radian/year.

Using a multitaper estimate with 3 and then 5 tapers on the differenced global mean temperature data gives Figure 7.14.

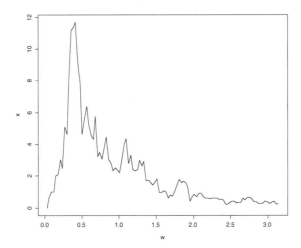

Fig. 7.12 *Beveridge spectrum using multitapers, average of 5*

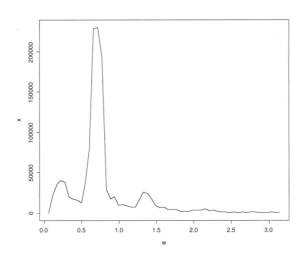

Fig. 7.13 *Lynx series spectrum using multitapers*

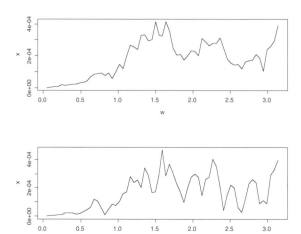

Fig. 7.14 *Differenced GMT series*

7.7 Parametric spectral estimates

Since any non-deterministic time series can be written in the form of a general linear model

$$X_t = \sum_{j=0}^{\infty} g_j \epsilon_{t-j}$$

then the spectrum can be regarded as a function of some time domain model. Indeed if we restrict ourselves to rational spectra of the form $|\phi(e^{-i\omega})/\theta(e^{-i\omega})|^2$ it seems only reasonable to estimate the parameters in the time domain. The most popular parametric estimates for the spectrum are the so called 'auto-regressive estimates'. These are obtained by fitting an autoregressive model $\phi(B)X_t = \epsilon_t$ and using as a spectral estimate

$$h(\omega) = \frac{\sigma^2}{|\phi(e^{-i\omega})|^2}$$

with the model estimates of σ and the autoregressive coefficients. A criterion is of course required to select the order of the autoregression and AIC is often used. Parzen (1974) suggested an alternative CAT criterion defined as follows. For each model, say of order m, based on a series of length N we have a residual variance σ^2 which we might write as v_m. Then

$$CAT(m) = \frac{1}{N} \sum_{j=1}^{m} \frac{1}{v_j} - \frac{1}{v_m}$$

with $CAT(0) = -1 - 1/N$. The model order we then select is that which minimizes the CAT.

Another possibility proposed by Akaike (1969) is to select the order which minimises the FPE (final prediction error) given by $(1 + (m + 1)/N)v_m$. There is little useful theory to help one choose a criterion. Jones (1976) suggests FPE underestimates the model order, while AIC is not consistent (statistically) as N tends to infinity. On balance AIC is most popular but there appears to be little difference between the criteria. Experience shows that the procedures tend to select AR orders in the range N/3 to N/2 for reasonable results. A relatively simple approach is to use the Yule–Walker equations and solve the system

$$\gamma(1) + \phi_1\gamma(0) + \phi_2\gamma(1) + \phi_3\gamma(2) + \cdots + \phi_p\gamma(p-1) = 0 \ (7.4)$$
$$\gamma(2) + \phi_1\gamma(1) + \phi_2\gamma(0) + \phi_3\gamma(1) + \cdots + \phi_p\gamma(p-2) = 0 \ (7.5)$$

$$\vdots \qquad\qquad (7.6)$$

$$\gamma(k) + \phi_1\gamma(k-1) + \phi_2\gamma(k-2) + \phi_3\gamma(k-3) + \cdots + \phi_p\gamma(k-p) = 0 \ (7.7)$$

for k=1,2, ... to obtain estimates for the model

$$X_t + \phi_1 X_{t-1} + \phi_2 X_{t-2} + \cdots + \phi_p X_{t-p} = \epsilon_t$$

and then to estimate σ^2 using the 'top' equation

$$\gamma(0) + \phi_1\gamma(1) + \cdots + \phi_p\gamma(p) = \sigma^2$$

The log of the Lynx series was chosen for analysis and the spectrum is displayed in Figure 7.15.

As can be seen the model order suggested by AIC is 11 which is confirmed by the other criteria. The least squares estimates give a set of model parameters and AR(11) model of the form

$$X_t - 1.139X_{t-1} + 0.508X_{t-2} - 0.213X_{t-3} + 0.270X_{t-4} - 0.113X_{t-5} + 0.124X_{t-6}$$
$$-0.068X_{t-7} + 0.040X_{t-8} - 0.134X_{t-9} - 0.185X_{t-10} + 0.311X_{t-11} = \epsilon_t$$

with residual variance 0.2262. The resulting spectrum is then

$$h(\omega) = 0.2262|1 - 1.139e^{-i\omega} + 0.508e^{-i2\omega} - 0.213e^{-i3\omega} \cdots + 0.311e^{-i11\omega}|^{-2}$$

which is quite spiked, the contrast with the nonparametric estimate is characteristic of the method. The Beveridge wheat price data give rise to an AR spectrum given in Figure 7.16, the AR process having order 8 from the AIC plot in Figure 7.17.

We are not always fortunate in finding a clear minimum for the AIC and the reader should be prepared to have to make a decision on the AR length on other grounds.

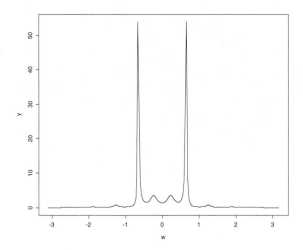

Fig. 7.15 *AR spectrum estimate for log of the Lynx series*

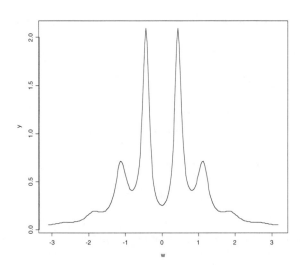

Fig. 7.16 *AR spectrum estimate for Beveridge series*

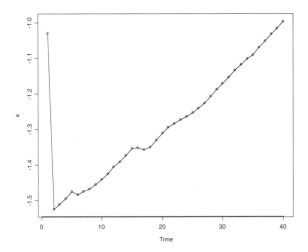

Fig. 7.17 *AIC plot for the Beveridge data*

While least squares is commonly used to estimate the parameters, however a variant known as 'maximum entropy' spectral estimation due to Burg (1968) minimises

$$\sum_{t=1}^{N-m} [X_t \phi X_{t+1} - \cdots - \phi_m X_{t+m}]^2 + [X_{t+m} - \phi_1 X_{t+m-1} - \cdots \phi_m X_t]^2$$

and is constrained to follow the Levinson recursion, essentially to satisfy the Yule–Walker equations. This latter condition is dropped in some variants. There is some evidence (Beamish and Priestley (1981)) that the unconstrained Burg method is rather better for cases with AR roots near the unit circle and for short series. The constrained version is known to suffer from 'line splitting' where a sharp peak in the spectrum is replaced by two or more closely spaced peaks. In addition the Burg estimate may give problems if the signal has a sinusoid as it is sensitive to the phase of the sinusoid and a frequency dependent bias may be introduced. There has been less work done on MA estimates of the spectrum of the form

$$h(\omega) = \frac{\sigma^2}{2\pi} |\Theta(e^{-i\omega})|^2$$

Brockwell and Davis (1990) give details and the distribution theory for such estimates. When deciding on a technique the central question is the use to which the spectral estimates are to be put. As we usually think of spectra as complementary to a time domain model this does imply a non-parametric approach is the sensible one. If the spectrum is the aim and a time domain development is of little interest then an AR estimate may be preferable.

8
Two or more series

It is clear that in many situations we have two or more time series which are of interest and we have to address the problem of modelling not just the individual series but also their interrelation. Thus in Figure 8.1 we see the catches of Mink and Muskrat in the Hudson Bay area for the years 1848 to 1911. It is inconceivable that these are not related in some way.

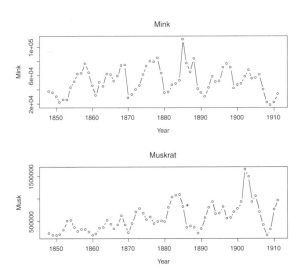

Fig. 8.1 *Annual Mink and Muskrat trappings from 1848*

In this chapter we make a start at setting up some of the ideas that are needed for modelling two or more series.

We know that there are autocorrelations for single series of the form

$$\gamma(k) = E[(X_t - \mu)(X_{t+k} - \mu)]$$

and it is an obvious step to define the cross correlations

$$\gamma_{xy}(k) = E[(X_t - \mu_x)(Y_{t+k} - \mu_y)] \qquad k = \ldots, -1, 0, 1, 2, 3, \ldots \qquad (8.1)$$

and in a similar way we have the cross correlations

$$\rho_{xy}(k) = \gamma_{xy}(k)/\sqrt{\gamma_{xx}(0)\gamma_{yy}(0)} \qquad k = \ldots, -1, 0, 1, 2, 3, \ldots \qquad (8.2)$$

The mediate implication of defining the correlation for two series is that *it is no longer symmetrical about zero*. This is reasonably obvious since there is no reason why two series should be most correlated when there is no lag between them. In mathematical terms

$$\gamma_{xy}(-k) = E[(X_t - \mu_x)(Y_{t-k} - \mu_y)]$$

putting $s = t - k$ we have

$$\gamma_{xy}(-k) = E[(X_{s+k} - \mu_x)(Y_s - \mu_y)] = E[(Y_s - \mu_y)(X_{s+k} - \mu_x)] = \gamma_{yx}(k).$$

This means that the cross covariances measure the direction (in time) of the association. As an example the cross correlations for the Mink and Muskrat series are plotted in Figure 8.2. There appears to be a large negative correlation at around lag 6 and a positive one at lag 13 but very little at lag zero!

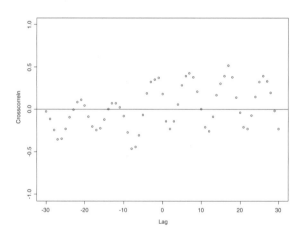

Fig. 8.2 *Mink and Muskrat crosscorrelations*

We extend the idea of stationarity in an obvious way. The two series are jointly (second order) stationary if each series is stationary and the cross covariances

$$\gamma_{xy}(t - s) = E[(X_t - \mu_x)(Y_t - \mu_y)]$$

depends only on $t - s$. Our main interest in pairs of series is often in finding the connection between them. Does X_t cause Y_t or vice versa? Perhaps there is a feedback so that the series are more subtly linked. Looking back at ARMA models one is tempted to think in terms of a system

$$a(B)X_t = b(B)Y_t + c(B)\epsilon_t \tag{8.3}$$

$$d(B)Y_t = e(B)X_t + f(B)\eta_t \tag{8.4}$$

Such complex systems are difficult to identify and to estimate so we lower our sights and look at some simpler forms.

8.1 Transfer function models

Suppose X_t and Y_t are stationary then a single input, single output output system is one where the series are related through

$$Y_t = v(B)X_t + \epsilon_t \tag{8.5}$$

where $\sum_{j=-\infty}^{\infty} v_j B^j$ is the transfer function of the system and ϵ_t is white noise, independent of the input series X_t. The model in 8.5 is usually known as a transfer function model or an ARMAX model. Notice the causal connection is one way, the X_t series causes the Y_t. This transfer function model is said to be *stable* if $\sum_{j=-\infty}^{\infty} |v_j| < \infty$, that is bounded inputs give rise to bounded outputs. If the transfer function $\sum_{j=-\infty}^{\infty} v_j B^j$ has zero coefficients for all negative j, that is $\sum_{j=0}^{\infty} v_j B^j$ then the system is known as a *causal* system.

We can attempt to identify causal systems if we make some assumptions about the transfer function. Suppose that it can be expressed as the ratio of two low order polynomials, say

$$v(B) = \frac{\omega(B)B^b}{\delta(B)} \tag{8.6}$$

Here $\omega(B)$ and $\delta(B)$ are low order polynomials in the backshift operator B. If we multiply out we have

$$[1 - \delta_1 B - \delta_2 B^2 \cdots - \delta_r B^r][v_0 + v_1 B + v_2 B^2 + \cdots]$$
$$= [\omega_0 - \omega_1 B - \omega_2 B^2 \cdots - \omega_s B^s]B^b$$

When we equate coefficients of B^j we have

$$v_j = 0 \qquad\qquad j < b \tag{8.7}$$

$$v = \sum_{i=1}^{r} v_{j-i}\delta_i + \omega_0 \qquad j = b \tag{8.8}$$

$$v = \sum_{i=1}^{r} v_{j-i}\delta_i - \omega_{j-b} \quad j = b+1, \ldots, b+s \tag{8.9}$$

$$v = \sum_{i=1}^{r} v_{j-i}\delta_i \qquad j > b+s \tag{8.10}$$

$$\tag{8.11}$$

It can then be shown that the coefficients of the impulse response satisfy the following

- v_j are zero for j less than b
- the coefficients from b to b+s-r have no real pattern
- $\delta_r(B)v_j = 0 \qquad j > b + s$

In theory we can deduce r, s and b by looking at the pattern of the impulse responses. All we need is some way of getting at these responses. If we take the equation 8.5 at time t+k

$$Y_{t+k} = v(B)X_{t+k} + \epsilon_{t+k} \tag{8.12}$$

and multiply by X_t we have, on taking expectations,

$$\gamma_{xy}(k) = v_0\gamma_{xx}(k) + v_1\gamma_{xx}(k-1) + \ldots \tag{8.13}$$

This is not very helpful unless the input X_t series is noise when the equation then gives

$$v_k = \sqrt{\frac{\gamma_{yy}(0)}{\gamma_{xx}(0)}}\rho_{xy}(k)$$

and the cross correlation gives a direct input to our transfer function. We then have a strategy.

1. Given our equation
$$Y_t = v(B)X_t + \epsilon_t \tag{8.14}$$

    and assuming that the input X_t is ARMA, say

    $$\phi_x(B)X_t = \theta_x(B)a_t$$

    We fit the model to obtain the residual series

    $$a_t = \theta_x^{-1}(B)\phi_x(B)X_t$$

2. Use the model in the previous step to obtain

    $$e_t = \theta_x^{-1}(B)\phi_x(B)Y_t$$

3. Now compute the cross correlations between the a_t and e_t series.
4. The required coefficients are

    $$\hat{v}_k = \sqrt{\frac{\gamma_{ee}(0)}{\gamma_{aa}(0)}}\rho_{ae}(k)$$

5. Now we identify b by looking at the patterns of the impulse response coefficients.

This gives us an initial stab at the transfer function. Once we have some idea then we can proceed to a direct likelihood procedure based on an extension of our state-space procedure.

In fact we can have a state space system of the form

$$X_t = H\alpha_t + \epsilon_t$$
$$\alpha_t = \phi\alpha_{t-1} + K\eta_t$$

where as in the single series case ϵ_t, η_t are vector noise streams, α_t is the state vector and X_t is the (vector) time series. The other matrices are parameters. If we extend the single series ideas we can find an expression for the likelihood and hence fit models, etc. Of course the devil is in the detail, the number of parameters becomes very large and the canonical forms rather more complex. The details are beyond our remit and we refer you to Janacek and Swift (1993) and Wei (1990).

8.2 The spectrum

In many ways it is simpler to deal with multiple series via the spectrum. Recall that for a single series X_t the spectrum $h_{xx}(\omega)$ is defined as

$$h_{xx}(\omega) = \frac{1}{2\pi} \sum_{s=-\infty}^{\infty} \gamma_{xx}(s)e^{is\omega} \tag{8.15}$$

Now we can in a similar way define the spectrum of the Y_t series

$$h_{yy}(\omega) = \frac{1}{2\pi} \sum_{s=-\infty}^{\infty} \gamma_{yy}(s)e^{is\omega}$$

and also the cross spectrum

$$h_{xy}(\omega) = \frac{1}{2\pi} \sum_{s=-\infty}^{\infty} \gamma_{xy}(s)e^{is\omega} \tag{8.16}$$

Notice that from previous results on the cross covariance

$$\gamma_{xy}(-\omega) = \gamma_{yx}(\omega)$$

we can see that

$$h_{yx}(\omega) = \frac{1}{2\pi} \sum_{s=-\infty}^{\infty} \gamma_{yx}(s)e^{is\omega} = \frac{1}{2\pi} \sum_{s=-\infty}^{\infty} \gamma_{xy}(-s)e^{is\omega} = \bar{h}_{xy}(\omega) \tag{8.17}$$

Thus for two series we have four spectra, indeed it is often useful to think of a matrix spectrum function

$$\mathbf{h}(\omega) = \begin{pmatrix} h_{xx}(\omega) & h_{xy}(\omega) \\ h_{yx}(\omega) & h_{yy}(\omega) \end{pmatrix}$$

or

$$\mathbf{h}(\omega) = \begin{pmatrix} h_{xx}(\omega) & h_{xy}(\omega) \\ \bar{h}_{xy}(\omega) & h_{yy}(\omega) \end{pmatrix}$$

This extends readily to cases where we have several series, but for the moment we look at the simplest case where we have just two series. Even so it means we have a complex function $h_{xy}(\omega)$. We can convert it to a real quantity as

$$h_{xy}(\omega) = c_{xy}(\omega) - iq_{xy}(\omega)$$

where

$$c_{xy}(\omega) = \frac{1}{2\pi} \sum_{s=-\infty}^{\infty} \gamma_{xy}(s)\cos(s\omega) \text{ and } q_{xy}(\omega) = \frac{1}{2\pi} \sum_{s=-\infty}^{\infty} \gamma_{xy}(s)\sin(s\omega)$$
(8.18)

The function $c_{xy}(\omega)$ is known as the cospectrum and $q_{xy}(\omega)$ is the quadrature spectrum.

An alternative is to write the complex quantity as

$$h_{xy}(\omega) = \alpha_{xy}(\omega)e^{i\phi_{xy}(\omega)}$$
(8.19)

Here $\alpha_{xy}(\omega)$ is the amplitude spectrum and ϕ_{xy} the phase spectrum. In fact the amplitude spectrum $\alpha_{xy}(\omega)$ measures the strength of the relationship between frequencies near ω in the two series. It is usual to normalize to get

$$K^2(\omega) = \frac{\alpha_{xy}(\omega)}{h_{xx}(\omega)h_{yy}(\omega)}$$
(8.20)

$K(\omega)$ is known as the coherency of the coherency spectrum and it is the correlation between the effects of the series at frequency ω. The other function one meets is the gain

$$G(\omega) = \frac{\alpha_{xy}(\omega)}{h_{xx}(\omega)}$$
(8.21)

Since we have just obtained a flock of functions we summarize them and their uses.

- The coherence or the squared coherency

$$K^2(\omega) = \frac{\alpha_{xy}(\omega)}{h_{xx}(\omega)h_{yy}(\omega)}$$

  measures the strength of the relationship between corresponding frequency components of two series in exactly the same way as a correlation coefficient

- The cross amplitude spectrum is an unscaled measure of association defined as

$$\alpha_{xy}(\omega) = |h_{xy}(\omega)|$$

- The gain

$$G(\omega) = \frac{\alpha_{xy}(\omega)}{h_{xx}(\omega)}$$

is the analogue of the regression of the frequency component at ω of the first series on the second.
- The lead or lag of this relationship is measured by the phase defined as

$$\phi_{xy} = \tan^{-1}\left(\frac{-q_{xy}}{c_{xy}}\right)$$

We digress into some mathematics to show how such results can be obtained.

8.2.1 The spectral representation

We can generalize the spectral representation we met in earlier chapters to the two series (and more) case. Suppose we have the X_t and Y_t series and their spectral representations

$$X_t = \int_{-\pi}^{\pi} e^{i\omega t} dZ_{xx}(\omega)$$

$$Y_t = \int_{-\pi}^{\pi} e^{i\omega t} dZ_{yy}(\omega)$$

Then we can show that the processes $Z_{xx}(\omega)$ and $Z_{yy}(\omega)$ are not only orthogonal so

$$E[dZ_{xx}(\omega)dZ_{xx}(\theta)] = \begin{cases} h_{xx}(\omega)d\omega & \text{if } \omega = \theta \\ 0 & \text{otherwise} \end{cases}$$

and

$$E[dZ_{yy}(\omega)dZ_{yy}(\theta)] = \begin{cases} h_{yy}(\omega)d\omega & \text{if } \omega = \theta \\ 0 & \text{otherwise} \end{cases}$$

but are also cross orthogonal

$$E[dZ_{xx}(\omega)dZ_{yy}(\theta)] = \begin{cases} h_{xy}(\omega)d\omega & \text{if } \omega = \theta \\ 0 & \text{otherwise} \end{cases}$$

This apparatus allows us to show the behaviour or the phase and the coherence for simple pairs of series. Suppose

$$Y_t = \beta X_{t-d} + \epsilon_t$$

where ϵ_t is noise uncorrelated with past X_t and Y_t. Then from the spectral representation

$$Y_t = \int_{-\pi}^{\pi} e^{i\omega t} dZ_{yy}(\omega) = \beta \int_{-\pi}^{\pi} e^{i\omega(t-d)} dZ_{xx}(\omega) + \int_{-\pi}^{\pi} e^{i\omega t} dZ_{\epsilon\epsilon}(\omega)$$

thus

$$dZ_{yy}(\omega) = e^{-di\omega}dZ_{xx}(\omega) + dZ_{\epsilon\epsilon}(\omega)$$

and multiplying by $dZ_{xx}(\omega)$ and taking expectations

$$h_{xy}(\omega) = e^{-id\omega}h_{xx}(\omega)$$

so the phase is

$$\phi(\omega) = -d\omega$$

So when there is a time delay the phase is a linear function of frequency.

Suppose now we have two transformed versions of our original series X_t and Y_t. Say

$$U_t = a(B)X_t \text{ and } V_t = b(B)Y_t$$

Where $a()$ abd $b()$ are polynomials in the shift operator B. It is clear that from our discussions of filter the (auto) spectra are

$$h_{uu}(\omega) = |a(e^{-i\omega})|^2 h_{xx}(\omega) \text{ and } h_{vv}(\omega) = |b(e^{-i\omega})|^2 h_{yy}(\omega)$$

but to get at the cross spectrum we need to do some simple calculations. Using the spectral representation we have

$$U_t = \int_{-\pi}^{\pi} e^{it\omega} dZ_{uu}(\omega) = \int_{-\pi}^{\pi} e^{it\omega} a(e^{-i\omega}) dZ_{xx}(\omega)$$

and

$$V_t = \int_{-\pi}^{\pi} e^{it\omega} dZ_{vv}(\omega) = \int_{-\pi}^{\pi} e^{it\omega} b(e^{-i\omega}) dZ_{yy}(\omega)$$

so from the cross orthogonal property we have

$$h_{uv}(\omega)d\omega = E[dZ_{uu}(\omega)d\bar{Z}_{vv}(\omega)]$$
$$= E[a(e^{-i\omega})b(e^{i\omega})dZ_{xx}(\omega)d\bar{Z}_{yy}(\omega)] = a(e^{-i\omega})b(e^{i\omega})h_{xy}d\omega$$

so

$$h_{uv}(\omega) = a(e^{-i\omega})b(e^{i\omega})h_{xy}$$

In consequence we have

$$|K(\omega)|^2 = \frac{|h_{uv}(\omega)|^2}{h_{uu}(\omega)h_{vv}(\omega)} =$$
$$\frac{|a(e^{-i\omega})b(e^{i\omega})h_{xy}(\omega)|^2}{|a(e^{-i\omega})|^2 h_{xx}|b(e^{-i\omega})|^2 h_{yy}} = \frac{|h_{xy}(\omega)|^2}{h_{xx}(\omega)h_{xx}(\omega)}$$

and we conclude that *the coherency is unchanged under such transformations*.

If the coherency is invariant under shifts in time then we can 'align' the series to get better estimates. You will recall that the spectrum is not symmetric about zero, neither is the cross covariance function. Since our univariate estimation ideas are based on this symmetry it is possible that there may be estimation problems, and there are! We can get some improvement, at least for coherence by shifting the series in time so that the maximum crosscorrelation occurs at lag zero. We know that this does not affect the value of the coherence so improvements in estimation are direct.

This alignment process is worth doing as it costs little and has some real benefits.

8.3 Applications

To illustrate these ideas we look at some examples. Two interesting series are the annual mean temperatures in the North and South hemispheres in 1856. As we can see from the plots in Figure 8.3 it seems that the individual series are non-stationary, the reader might like to check this by examining the autocorrelations. They are however linked as we see when we plot the series against each other as in Figure 8.4.

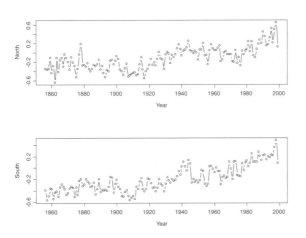

Fig. 8.3 *Annual temperatures in the Northern (top) and Southern hemispheres*

Plotting the crosscorrelations of the differenced series (Figure 8.5) shows us the maximum correlation occurs at lag zero, and there are two other large values at lag 2 and 18. Why there should be a lag of 2 years is something we will leave to the climatologists. If we consider the cross spectrum we see, in Figure 8.6 that the univariate spectrum for the differenced southern hemisphere temperatures has a peak and a large amount of high frequency power. There is a similar peak in the northern hemisphere temperatures but there is not an equivalent amount of power at high frequencies. The coherency is high as we might expect, except

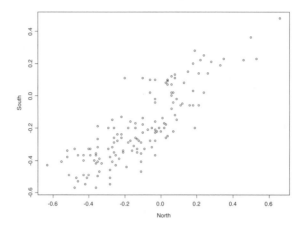

Fig. 8.4 *Temperature differences North vs.South*

at high frequencies and the phase plot does not appear to exhibit any real trend. One interesting phenomenon which affects the weather is the 'El Nino' or the Southern oscillation. The value of the index for the years 1866–1997 is shown in Figure 8.5.

The question is whether this index can be related to climate changes in the southern hemisphere. The crosscorrelation between the index and temperature *change* is shown in Figure 8.6.

Giving lags at -1 and 1. The cross spectrum shown in Figure 8.7 is also less than informative. Apart from three peaks the coherency is low and the phase looks flat.

We started with the mink muskrat data so it sees appropriate to finish with this data set. The crosscorrelations in Figure 8.8 look rather odd. The sinusoidal effect is related to the cyclic behaviour of the individual series which can be seen by examining the autocorrelations, see Figure 8.8.

Filtering the 10-year seasonal gives a new view in the crosscorrelations and the crosspectra (Figures 8.9, 8.10 and 8.11).

There is clearly some connection between the series, we have positive crosscorrelations which are meaningful. The coherency is rather jagged, perhaps more smoothing would be helpful, but there seems to be some indication of a low frequency link. There is also some suggestion of a decaying phase indicating a lag between the series.

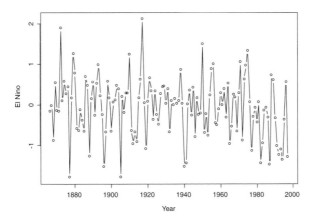

Fig. 8.5 *Annual El Nino index 1866-1997*

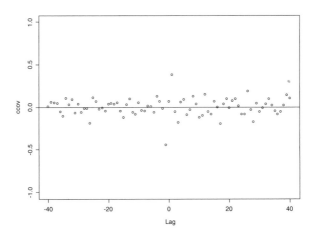

Fig. 8.6 *Cross correlation El Nino index and Southern hemisphere temperature change*

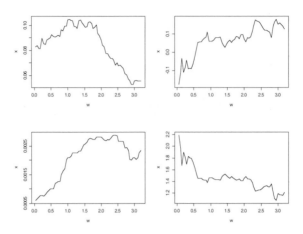

Fig. 8.7 *Cross spectrum El Nino index and Southern hemisphere temperature change*

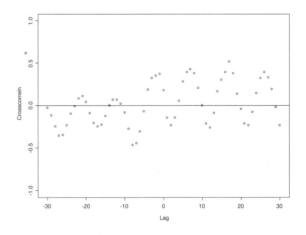

Fig. 8.8 *Crosscorrelations for mink and muskrat*

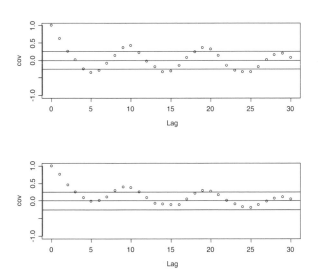

Fig. 8.9 *Autocorrelations for mink and muskrat*

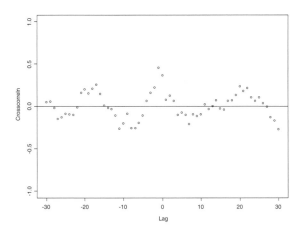

Fig. 8.10 *Cross correlations for filtered mink and muskrat*

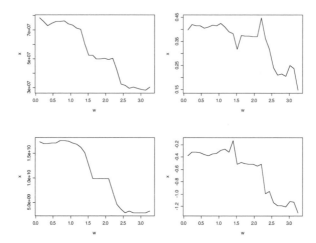

Fig. 8.11 *Spectra for filtered mink and muskrat, top autospectra, bottom coherency and phase*

It should be obvious to the reader that the same ideas we discussed in earlier chapters can be extended to vector series, that is multiple time series. We have to stop somewhere and as we cannot provide the software support, bivariate series are as far as we feel we can go. The multiple case is more complex but the major concepts have all been discussed earlier in this book.

We leave the reader with a simple message – try out the methods on the data provided – or even better, on your own data sets. There is very often not a neat and simple answer but patience and persistence will pay dividends.

9
The R language

9.1 Time series in R

When we began this project there were no time series functions in R and we wrote
our own. They are all written in R and are thus rather slower than similar functions
written in C and complied. They do have two advantages.

1. They are transparent in that the user can just type the function name to get
   a listing of the function
2. They are easily modified

Thus for example if we are interested in plotting time series the function tsplot is
provided. When `tsplot` is typed we have the listing

```
> tsplot
function (x)
{
    # time series plotter
    time <- seq(1, length(x))
    plot(time, x)
    lines(time, x)
}
```

If we edit the function by removing the line `lines(time, x)` then the plot
becomes just points rather than points and lines. A brief catalogue of functions
follows. The functions are provided to allow the reader to experiment with ideas
and recipes in the text. They have been tested but there are no guarantees and the
author is not responsible for any loss or damage which might arise from their use.
They are all available from the web address

`www.mth.uea.ac.uk/ h200/book/`

together with some documentation. The list of functions at the URL given will
contain rather more than the list below

9.1.1 The Elements of R

Once you are in R:
To get help on a topic, say print type

```
?print
```

To quit R type

```
q( )
```

Beware both UNIX and R are case sensitive so GLM and glm are not the same.

Data

Data are held in scalars, vectors, matrices and data frames. The assignment to a scalar is by the left arrow operator

```
x<- 4
x<-y^3+99
```

It is simplest to read data from a keyboard into a vector using the scan( ) function, thus

```
y <- scan( ) (return)
```

will prompt you for a number. Hitting (return) gives a new prompt (return) followed by (return) ends the data input.
So

```
 x<-scan ( )
1
2
3
4
>
>
```

- a new line stops input.
An alternative is to use c

```
 x<-c (1,2,3,4)
```

but I would stick with scan.
Vectors can be used in arithmetic so

```
V <-x * x - 7 / x
```

generates a new vector V. There are the usual functions and arithmetic operators $+, *, /, -,$ log, exp . tan, $sqrt$. Other useful functions are

```
max(x)
min(x)
length(x)  -  number of elements in   x
sum(x)
mean(x)
sort(x)  -  elements in increasing order
r norm(x)  -  produces a set of random standard normal deriva-
tives of the
    same size as   x
```

Sequences

We can use shortcuts to generate regular sequences

```
x <-c(1,2,  ...,  3c) is x<-1:30
```

or other examples

```
2*1:15
  30:1
```

Note: : has highest priority. The function seg( ) is more general for example

```
S3 <-seg(-5,5, by = 0.2)
```

rep( ) can also be useful, viz

```
S4<-rep(x, times - 5)
```

Factors

A special vector we often use is the factor used to specify a discrete classification

```
state <- c("big, "small", "tiny")
```

```
state<-factor (state)
```

```
treat <- c(1,2,3,1,2,3,1,2,3)
treat<-factor (treat)
```

Matrices

R quite a lot about matrices so it is useful to know

```
dim(z) <- c(2,5)
```

lets R recognize z as a 2 by 5 matrix.

Data Frames

A data frame is a particularly useful object. The simplest way to think of it as a matrix with columns corresponding to variables and rows as observations. I use

```
lentil<-read.table( file=name,header=TRUE)
```

For example lentil might be:

```
u    v
1    5
2    8
3    11
```

We can refer to the columns as

```
lentil$u
lentil v
```

This can be irritating so we use `attach (lentil)`. This means we can use the column names in the frame. As we said above t is best to read data from files into a data frame by

```
newt<- read.table(file="file")
```

If the columns in the file have names then you need to tell R that there are headers viz.

```
 newt<-read.table("file", header = T)
```

There are many data sets available in R. See ? `data.`

Graphics

```
plot (x,y) plots y against x
plot (f,g) where f is a factor produces boxplets for each
level of f
plot (df) df is a data frame
plot (y~expr) plots y against every object in expr
```

Of course your best bet is to read the R documentation and to use my time series functions.

9.2 General functions

tsplot(series)

Given a series this function plots the points and joins them with lines

splot(series)

Given a series this function plots the points against frequencies.

acf(series,k)

Given a series the function returns the first k covariances in cov. Note cov[j] is $\gamma(j+1)$ the $j+1$ covariance.

acfplot(series,k)

Given a series the function returns the first k autocorrelations in cov. Note cov[j] is $\rho(j+1)$ the $j+1$ autocorrelation. The Autocorrelations are plotted and bands drawn to indicate the significant values.

pacf(cov,k,pt)

Given a series the function returns the first k partial autocorrelations in cov. If pt is set to T then the partial autocorrelations are plotted. Note cov[j] is $\phi(j+1)$ the $j+1$ partial autocorrelation.

9.3 Smoothing

window(series,w)

Given a series and a vector of weights the function returns the smoothed series. Note we assume that the w is of odd length but w is normalized to weight 1.

windowlook(x,w)

This function uses window and returns a smoothed series as for window. In addition the original and smoothed series are plotted.

msmooth(x,k)

Smoothes a series a using a median smoother. Successive segments of length k are used to get the median.

msmoothlook(x,k)

Median smoother but provides plots of the original and the smoothed series.

9.4 Simulation

ma(theta,n)

Generates a moving average of length n. The coefficients are in theta.

ar(phi,n)

Generates an autoregressive series of length n. The coefficients are in phi.

arma(phi,theta,n)

Generates an arma(p,q) model series. The coefficients ate in phi and theta.

9.5 Identification

corner(cov)

This is an implementation of the corner method to identify p and q for arma models. See Gourieroux *et al.* (1980). The cov parameter must hold at least 25 autocorrelations.

normcorner(cov)

A normalized version of corner

stepar(cov,nmax,nseries)

This function fits AR models of increasing length. It requires the autocorrelations in cov, a max order in max and the length of the series in nseries. The criteria are AIC and BIC. This is verbose so the user can see the outcome

9.6 Exponential smoothing

ssex(x,n,a)

Gives an exponentially smoothed sequence based on a series x with smoothing parameter a.

tec1(x)

Plots the mean square error against smoothing parameter for single parameter exponential smoothing.

sex2(x, a, b, k)

Produces mse for a two prameter exponetial smoothing model for series x. Uses first k values for the mean square error.

eforecast2(x, k)

Minimizes the mse for a two parameter exponential smoothing model. Uses the first k values of x.

look(x,k)

Gives a table of mse for a two parameter exponential smoothing model.

9.7   Estimation

kalmanf(data, n, d, phi, k)

Computes the likelihood for a state space model using Kalman filter recursions. Output is the likelihood. Designed for ARMA models d=max(p,q) phi k (see Janacek and Swift (1993)).

kalmanr(data, n, d, phi, k)

As Kalman f but ouput is a list with prediction errors, etc.

kalmanq(data, n, a, bc, h, sigma, var)

A more general Kalman function.

setarma(vp)

Sets up the state space model for kalman routines. vp contains the ar and ma coeffs as vp<- c(phi,theta).

allarma(vp)

Sets up the state space model for kalman routines. vp contains the ar and ma coeffs as vp<- c(phi,theta). This is used when we have a minimum.

9.8   Frequency domain

dft smoother

Smoothes a series by setting high frequencies to zero.

pgram

Given a series computes the periodogram.

specm

Computes the spectrum by smoothing the periodogram using a rectangular window.

specp

Computes the spectrum by smoothing the periodogram using a Parzen window.

mtaper

Computes the spectrum using mutitapers.

non-parametric spectral estimate

stepar

Finds the best AR model using AIC.

pspec

Computes the spectrum given an AR model.

9.9   Utility functions

odd(x)

Is zero if x is even otherwise it takes the value 1. Note it thinks zero is even.

tri(cov, i, j)

A function called by the corner function.

crossprod(a,b)

This function returns the sum a(i)*b(i).

qchi(p,dof)

Chi squared quantiles.

9.10   Bivariate functions

crossspec2(x, y,k, pt = T)

Computes and plots the univariate specta for two series, the coherency and phase.

ksmoothall(x,y,h)

A not very clever kernel smoother which smoothey y over x using a bandwidth h. Kernel is given in function epan. The functions ksmooth, kern, and epan are subfunctions.

crosscorr(x,y,k,pt=T)

Computes and plots cross correlations.

References

Anderson, TW *The statistical analysis of time series*, Wiley, New York, 1971.

Akaike, H Fitting autoregressive models for prediction, *Annals of the Inst of Statistical Maths* **21** (1969), 243–7.

Beamish, N and Priestley, MB A study of autoregressive estimation, *Applied Statistics* **30** (1981), 41–58.

Bloomfield, P *Fourier analysis of time series*, Wiley, New York, 1976.

Box, GEP and Jenkins, G *Time series analysis: Forecasting and control*, Holden-Day, San Francisco (Reprinted 1976), 1970.

Box, GEP and Pierce, DA Distribution of residual autocorrelations in autoregressive integrated moving average models, *J Am Stat Assoc* **65** (1970b), 1509–26.

Brockwell, PJ and Davis, RA *Time series: Theory and methods*, Springer, New York, Second edition, 1990.

Brown, RG *Statistical forecasting for inventory control*, McGraw-Hill, New York, 1959.

Brown, RG *Smoothing forecasting and prediction of discrete time series*, Prentice-Hall, New Jersey, 1963.

Burg, JP *Maximum entropy spectral analysis reprinted: Modern spectrum analysis*, (ed. D G Childers), IEEE Press, New York, 1968.

Camina, AR and Janacek, GJ *Mathematics for seismic data processing and interpretation*, Graham and Trotman, London, 1984.

Chatfield, C The holt winters forecasting procedure, *Applied Statistics* **27(3)** (1978), 264–79.

Chatfield, C and Prothero, DL Box-jenkins seasonal forecasting: problems in a case study, *Journal Royal Statistical Society* A **136** (1973), 259–336.

Cooley, J and Tukey, J An algorithm for the machine calculation of complex Fourier series, *Math Comp* **19** (1965), 297–301.

Fisher, R Tests of significance in harmonic analysis, *Proc Roy Soc London Ser* **125** (1929), 54–9.

Gourieroux, C, Beguin, JM, and Montfort, A Indentification of an arima process: the corner method, *Time Series* (T Anderson, ed.), North Holland, Amsterdam, 1980.

Gourieroux, C and Montfort, A *Time series and dynamic models*, Cambridge University Press, Cambridge, 1997.

Hardle, W *Applied nonparametric regression*, Cambridge University Press, Cambridge, 1990.

Harrison, PJ and Stevens, CF Bayesian forecasting, *Journal Royal Statistical Society* **38** (1976), 3.

Harvey, AC *Time series models*, Philip Allan, Oxford, 1981.

Harvey, AC *Forecasting structural time series models and the kalman filter*, Cambridge University Press, Cambridge, 1989.

Holt, CC *Forecasting seasonals and trends by exponentially weighted moving averages*, ONR Research Memorandum, Carnigie Institute 52, 1957.

Janacek, GJ and Swift, AL *Time series*, Ellis Horwood, Chichester, 1993.

Jones, RH Autoregressive order selection, *Geophysics* **41** (1976), 1771–3.

Kalman, RE A new approach to linear filtering, *Trans ASME J of Basic Engineering* **82** (1960), 35–45.

Koopmans, L *The spectral analysis of time series*, Academic Press, New York, 1974.

Mallows, CL Some theory of nonlinear smoothers, *Annals of Statistics* **8** (1976), 695–715.

Neave, H Spectral analysis of stationary time series using initially scarce data, *Biometrika* **57** (1970), 111–22.

Parzen, E Some recent advances in time series modelling, *IEEE Trans Automatic Control* **19** (1974), 723–30.

Percival, DB and Waldren, AT *Spectral analysis for physical applications*, Cambridge University Press, Cambridge, 1993.

Priestley, MB *Spectral analysis and time series*, Volumes I and 2, Academic Press, New York, 1981.

Riedel, KS and Sidorenko, S Minimum bias multitaper spectral estimation, *IEEE Transactions of signal processing* **43** (1993), 188–95.

Shumway, RH and Stoffer, DS An approach to time series smoothing and forecasting using the em algorithm, *J. Time series Anal.* **3** (1982), 253–64.

Teukolsky, SA, Press, WH, Flannery, BP, and Vetterling, WT *Numerical recipes*, Cambridge University Press, Cambridge, 1986.

Thomson, DJ Spectrum estimation and harmonic analysis, *Proceedings of the IEEE* **70** (1982), 1055–96.

Velleman, PF Definition and comparison of robust nonlinear data smoothing algorithms, *Journal of the American Statistical Assn.* **75** (1980), 609–15.

Wei, WS *Time series analysis, univariate and multivariate methods*, Addison Wesley, New York, 1990.

West, M and Harrison, J *Bayesian forecasting and dynamic methods*, Springer-Verlag, New York, 1989.

Whittle, P *Prediction and regulation by linear least squares methods*, English Universities Press, (Reprinted, Blackwell, Oxford, 1994), 1963.

Wilks, SS *Mathematical statistics*, Wiley, New York, 1962.

Winters, PR Forecasting sales by exponentially weighted moving averages, *Management Science* **6** (1960), 324–42.

Index